Particulate Interactions in Dry Powder Formulations for Inhalation

Particulate Interactions in Dry Powder Formulations for Inhalation

Xian Ming Zeng
Gary P. Martin
Christopher Marriott
Department of Pharmacy,
King's College London

Routledge
Taylor & Francis Group

LONDON AND NEW YORK

The publishers have made every effort to contact authors/copyright holders of works reprinted in *Particulate Interactions in Dry Powder Formulations for Inhalation*. This has not been possible in every case, however, and we would welcome correspondence from those individuals/companies we have been unable to trace.

First published 2001
by Taylor & Francis

2 Park Square, Milton Park, Abingdon, Oxfordshire OX14 4RN
52 Vanderbilt Avenue, New York, NY 10017

Routledge is an imprint of the Taylor & Francis Group, an informa business

First issued in paperback 2019

Typeset in 9/14pt Stone Serif by
Wearset, Boldon, Tyne and Wear

British Library Cataloguing in Publication Data
A catalogue record for this book is available from the British Library

Library of Congress Cataloging in Publication Data
Zeng, Xian Ming, 1963–
    Particulate interactions in dry powder formulations of inhalation/Xian Ming Zeng, Gary P. Martin, Christopher Marriott.
        p. cm.
    Includes bibliographical references and index.
    ISBN 0-7484-0960-2 (alk. paper)
    1. Powders (Pharmacy) 2. Aerosol therapy. I. Martin, Gary P., 1954– II. Marriott, Christopher, 1944– III. Title.
    RS201.P8 Z46 2000
    615.8'36–dc21
                                                    00–044702

ISBN: 978-0-7484-0960-0 (hbk)
ISBN: 978-0-367-39797-5 (pbk)

Contents

Interparticulate Forces

Contents

1.1 INTRODUCTION

In the western world most drugs are administered as solid dosage forms which means that, with a few exceptions, they will exist in the powdered state at some stage of their manufacture. Therefore, it is important for the pharmaceutical developer to have a comprehensive understanding of the properties of powdered materials. Most regulatory bodies have focused primarily on issues of safety, quality and efficacy, which in turn has led to an overriding interest in all aspects of chemical purity. However, it has been recognised also that the physical state of a solid formulation may sometimes be as important as the chemical structure of the drug since its *in vivo* disposition is often determined by many of its physical properties. For example, absorption of a drug entity of low aqueous solubility from the gastrointestinal tract is often dependent upon the particle size of the drug in the formulations. Polymorphism is another important factor in determining drug disposition both *in vitro* and *in vivo*. It is widely known for instance that amorphous drugs have a higher molecular mobility than crystalline drugs and hence, the former dissolve more readily in aqueous solutions, often resulting in a higher rate of absorption than when the drug is presented in the latter state. However, the amorphous state is thermodynamically unstable and it will eventually transform to a more stable crystalline form under normal conditions. Such a transformation is known to result in physical and even chemical changes in the drug formulation, which may eventually lead to an alteration in the pharmacological effects of the formulation.

The need for physical characterisation becomes even more crucial when excipients are used in formulations. With the rapid development in the material sciences in recent years, a wide variety of pharmaceutical excipients have been used and these will undoubtedly have a concomitant range of physico-chemical properties. To improve the safety and efficacy of drug formulations, the physical characterisation of drugs, excipients, and blends of the two is gradually becoming a part of preformulation studies. Various spectroscopic methods (including ultraviolet/visible, diffuse reflectance spectroscopy, vibrational spectroscopy and magnetic resonance spectrometry), optical and electron microscopy, X-ray powder diffractometry and thermal methods of analysis, etc. have been used to characterise the physical states of many pharmaceutically-active therapeutic agents (Brittain, 1995). A systematic approach to the physical characterisation of pharmaceutical solids has been outlined (Brittain *et al.*, 1991) and the physical properties classified as being associated with the molecular level, the particulate level and the bulk level. The use of these techniques has undoubtedly promoted a scientific and systematic approach to the development of pharmaceutical solid dosage forms and has provided formulation scientists with invaluable information that has optimised therapeutic effects and minimised toxic reactions.

However, many associated difficulties still exist with the study of solids in the powdered state. Even current technological advances have not produced a comprehensive or exhaustive insight into the characterisation of solids incorporated within dosage forms. This is particularly true in the case of the particulate interactions between powders. The vast majority of drugs, when isolated, exist as either crystalline or amorphous solids. They are normally milled (comminuted), mixed with other inactive excipients (such a term may not always be accurate since some excipients do have biological activities) and may be finally filled into capsules or compacted to form tablets. Other solids, after comminution, may be suspended in a suitable medium to produce dosage forms such as suspensions. Particles in powdered dosage forms interact with each other throughout all the powder handling process from comminution, through mixing and compaction, to storage, due to the ubiquitous attraction and/or repulsion forces. Particulate interaction has been actively researched in other scientific and technological fields, such as powder technology and in the semiconductor industry, but this has not been the case with pharmaceutical solids and much of the pertinent work is scattered throughout the literature. The importance of particulate interaction in pharmaceutical solid dosage forms must not be underestimated since many bulk properties of powders, such as flowability, mixing, deaggregation, dispersion, compression and even drug dissolution, may be affected by the particle–particle interaction between the solids (Figure 1.1). In order to provide a more systematic insight into these interactions in pharmaceutical solid dosage forms, it is necessary to introduce some of the basic principles generated from other scientific fields. The emphasis of this work will be placed on the type of particulate interactions and factors that may affect them together with their relevance to

■ CHAPTER 1 ■

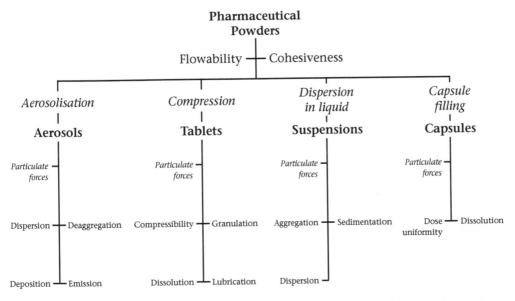

Figure 1.1: Flow chart showing some of the properties of pharmaceutical solid dosage forms which may be affected by particulate forces.

pharmaceutical solid dosage forms. Special focus will be placed on their effects on the powder flowability, mixing, drug delivery from aerosols, tablets, capsules and suspensions.

Particle interactions occur as a consequence of so-called long range attractive forces. The forces arising from the covalent bonds that result in the formation of molecules from atoms are not covered by this definition. Covalent forces bring two atoms together to form a molecule which has completely different properties to that of the component atoms. Such bonds are extremely strong with an energy of formation often greater than 40 kJ mol^{-1} (300 to 700 kJ mol^{-1} being typical). Particle interactions, however, are due to much weaker physical bonding usually with energies of bonding less than 40 kJ mol^{-1}. These forces are thus weaker than chemical bonds but their influence extends over greater distances and they are thus termed, 'long range forces'.

Particulate interactions can be broadly classified into two classes, namely cohesive and adhesive (Zimon, 1982). Cohesion usually refers to interactions between particles of the same chemical structure and of similar particle size. Adhesion refers to interactions between particles of different materials. Thus, a powder is said to be cohesive when the component particles tend to stick to each other and the interaction forces between these particles are termed cohesive forces. On the other hand, a particle is adhesive if it readily adheres to the surface of an object of larger dimensions. Similarly, the forces between a particle and such a surface are defined as the adhesive forces. Either cohesion or adhesion becomes significant when gravitational forces acting upon these particles become negligible; that is, when the dimensions of the particulate materials become smaller than 10 μm (Visser, 1995). Many pharmaceutical solids fall within this size range and hence, either cohesion or adhesion is commonly encountered in drug formulations.

Particulate interactions may be a result of a number of concurrently acting forces or mechanisms, van der Waals, electrostatic, capillary forces, solid bridging or mechanical interlocking being typical. Since pharmaceutical solids are often insulators, particles are likely to carry a charge during the powder handling process. Charged particles will exert so-called electrostatic forces (Coulombic) on other particles. However, uncharged particles also interact with each other due to dispersion forces (van der Waals forces). They are very complicated but may be understood by imagining an instantaneous picture of molecules possessing different electronic configurations, giving them a dipolar (having both a negative and positive pole) character. This temporary situation will result in local imbalances in charge, leading to transient induced dipoles. Consequently, the molecules attract each other. Dispersion forces may exert their influence over a range in the order of 10 nm and have a typical strength, for van der Waals bonds, of around 1 kJ mol^{-1}. When water (or other liquid) condensation occurs on the solid–solid interfaces, then particulate interaction due to capillary forces arises. Such forces, resulting from the surface tension of the adsorbed liquid layer, may dominate over other forces if sufficient liquid is condensed on the solid–solid interface. Finally, pharmaceutical solids are often irregularly shaped with rough surfaces. Mechanical interlocking between particles is a common occurrence once they are in contact, especially after compaction. Thus, mechanical interlocking can also be an important contributor to overall particulate interactions.

1.2 VAN DER WAALS FORCES

The major forces between uncharged solid particles are named after the Dutch physicist van der Waals, who pointed out more than a century ago that the deviations from the ideal-gas law at high pressures could be explained by assuming that the molecules in a gas attract each other (van der Waals, 1873). More than half a century later, by employing the then newly developed quantum chemistry, London was able to quantify van der Waals' hypothesis (London, 1930) and hence, the forces are also called van der Waals–London forces. According to London's theory, the energy of interaction between two molecules, V, can be given as the following expression:

$$V = - \frac{3}{4} h\nu_0 \frac{\alpha^2}{d^6} \tag{1.1}$$

where α is the polarisability, h is Planck's constant, ν_0 the characteristic frequency and d the separation distance between the two objects. Since ν_0 is found in the ultraviolet region of the absorption spectra and plays a key role in optical dispersion, van der Waals–London forces are also termed dispersion forces.

By assuming the additivity of the energy of interaction by dispersion forces, Hamaker (1937) was able to calculate the interaction energy between two solid objects simply by summarising all the possible individual molecular interactions and the energy changes of interaction between macroscopic bodies:

$$E = - \int_{V_1} \int_{V_2} \frac{q_2 q_2 \, dv_1 \, dv_2 \, \lambda_{1,2}^d}{r^6} \tag{1.2}$$

where E is the total energy of interaction; v_1, v_2 are the total volumes of particles, 1 and 2; q_1, q_2 are the number of atoms per unit volume of particles 1 and 2; r is the distance between volume element 1 and 2.

The constant of dispersion ($\lambda_{1,2}^d$) in equation (1.2) can be calculated from the following equation:

$$\lambda_{1,2}^d = - \frac{3}{4} h f a^2 \tag{1.3}$$

where h is the Planok's constant; f is the vibration frequency of the interacting electronic oscillators; a is the polarisability of molecules.

Van der Waals forces, F, between two ideally smooth spheres of diameters d_1 and d_2, separated by a distance r in vacuum can be expressed as:

$$F = \frac{A}{12r^2}\left(\frac{d_1 d_2}{d_1 + d_2}\right)$$ (1.4)

The interaction forces between two flat plates per unit of surface area, P_A, are given as:

$$P_A = \frac{A}{6r^3 \pi}$$ (1.5)

where A is Hamaker's constant which is given by:

$$A = \pi^2 q_1 q_2 \lambda_{1,2}^d$$ (1.6)

If a particle with a diameter d_1 is separated by a distance r in vacuum from a plane surface, the adhesive force of a sphere to a plane composed of the same molecules due to van der Waals forces is given as:

$$F = \frac{A_{11} d_1}{6r^2}$$ (1.7)

The adhesion of a sphere (1) to a plane surface (2) composed of different molecules can be calculated as:

$$F = \frac{\sqrt{A_{11}A_{22}}}{6r^2} d_1$$ (1.8)

where A_{11} and A_{22} are the Hamaker constants for the fine particle and the plane surface, respectively; d_1 is the diameter of the particle and r is the separation distance between the particle and the plane surface.

Although the above equations have been used to estimate the interaction forces between two macroscopic objects, the additivity of the interactions is subject to question since the molecules at the surface of the particle are thought to have a so-called screening effect on the interactions between molecules located in the inner layers of the particle. Hence, the interaction between solid bodies was found to be determined solely by the electromagnetic properties of the material contained in the surface layer of the same thickness as the separation distance (Langbein, 1969). In order to circumvent Hamaker's assumption of the additivity of molecular interactions, both Lifshitz (1956) and Dzyaloshinskii and co-workers (1960), developed a theory for the interaction energy between solid bodies *in vacuo* and in a liquid medium, using bulk material properties only. However, the equations these authors obtained were identical to those given above except that the Hamaker constant (A) could now be linked to measurable bulk material properties, e.g. the optical characteristics from visible light to the far ultraviolet and beyond.

The Lifshitz–van der Waals constant, $h\varpi$, is related to the Hamaker constant by:

$$A = \frac{3}{4\pi} h\varpi \qquad (1.9)$$

Lifshitz–van der Waals constants of many materials in vacuum and in water have been calculated (Visser, 1995) and the more electron-dense the material, as in the case of pure metals, the larger the constants will be. Immersion in a third medium will reduce the van der Waals interaction.

If interactions occur between particles of different materials, the Hamaker or Lifshitz–van der Waals constant A_{12} can be approximated by taking the geometric means of the individual coefficients A_{11} and A_{22}:

$$A_{12} = \sqrt{A_{11}A_{22}} \qquad (1.10)$$

For the same combination in a third medium, the constant A_{132} becomes:

$$A_{132} = (\sqrt{A_{11}} - \sqrt{A_{33}})(\sqrt{A_{22}} - \sqrt{A_{33}}) \qquad (1.11)$$

When $A_{11} < A_{33} < A_{22}$ or $A_{11} > A_{33} > A_{22}$, then $A_{132} < 0$.
Otherwise, $A_{132} > 0$.

Thus, A_{132} can be positive, indicating that the two objects attract each other or it can be negative, indicating that the two objects repel each other in the medium. As can be seen from the data presented in Table 1.1, immersion in water leads to a substantial reduction in the van der Waals attraction.

TABLE 1.1
Lifshitz–van der Waals coefficients, $h\varpi$, of various materials in vacuum and in water (Visser, 1989)

Combination	$h\varpi$ (eV)	
	Vacuum	Water
Au–Au	14.30	9.85
Ag–Ag	11.70	7.76
Cu–Cu	8.03	4.68
C–C (diamond)	8.60	3.95
Si–Si	6.76	3.49
Ge–Ge	8.36	4.66
MgO–MgO	3.03	0.47
KCl–KCl	1.75	0.12
Al_2O_3–Al_2O_3	4.68	1.16
CdS–CdS	4.38	1.37
H_2O–H_2O	1.43	–
Polystyrene–polystyrene	1.91	0.11

1.3 ELECTROSTATIC FORCES

In the gaseous environment, apart from the electrodynamic forces of van der Waals type, electrostatic forces can also contribute substantially to particulate interactions. Electrostatic charges can arise from the contact of an uncharged particle with either a negatively or positively charged particle, owing to the transfer of electrons and ions amongst particles. Charge may also result from the contact of two uncharged particles that have different work functions. This latter parameter is defined as the minimum energy required to move the weakest bound electron from a particle surface to infinity. Apart from contact, sliding and friction also produce electrical charges and therefore the term, 'triboelectrification' is often used to describe the process.

When two charged particles are brought together, they will experience attraction or repulsion depending on their electrical charge signs. In the simplest case of two-point charges, the magnitude of the electrostatic forces (F_{el}) between them can be described by Coulomb's law:

$$F_{el} = \frac{q_1 q_2}{4\pi\epsilon d^2}$$ (1.12)

where q_1 and q_2 are the electrical charges on the two particles respectively, ϵ is the permitivity and d is the distance of separation.

If a particle of charge, q, approaches an uncharged particle of radius r, the first particle will induce an image charge on the second particle. The electrostatic force (F_c) between these particles is then calculated as:

$$F_c = \frac{q^2\left(1 - \dfrac{d}{\sqrt{r^2 + d^2}}\right)}{16\pi\epsilon d^2}$$ (1.13)

where d is the distance between the charged and uncharged particles.

When two solids of different work functions in vacuum are separated by a small distance (d), electrons from the material with the lower work function are transferred to the material with the higher one. The two work functions will eventually reach the same level and thus an equilibrium will be established. Then, the forces of attraction (F_w) can be calculated using the expression:

$$F_w = \pi\epsilon r \frac{(\Delta U)^2}{d}$$ (1.14)

where ΔU is the potential difference arising from the difference in work functions and r is the radius of the interactive particles.

The role of electrostatic forces in particle–surface interaction is similar to that in particle–particle cohesion. When a charged particle approaches a flat earthed conducting surface, the particle will also induce an image charge on the plane surface. The adhesion force of the charged particle to the surface due to electrostatic forces can be calculated according to equation (1.13) when $r \to \infty$:

$$F_e = \frac{q^2}{16\pi\epsilon d^2} \tag{1.15}$$

and $\quad q = e(N_d - N_a)S \tag{1.16}$

where e is the charge of an electron, N_d and N_a are the surface concentrations of the donor centres and acceptor centres respectively and S is the contact area between the particle and surface. Hence, electrostatic forces may be increased or reduced by changing the electrical properties of either the particle or surface, and these are largely dependent upon the chemical structures and physical states of the materials concerned. Some of the possible molecular groups have been ranked in a donor–acceptor series (Muller *et al.*, 1980):

Donor $\quad -NH_4 > -OH > -OR > -COOR > -CH_3 > C_6H_5 > -$halogen $>$ $=CO > -CN$ Acceptor

Each successive member is an acceptor with respect to the preceding member, which acts as a donor. Therefore, materials will have different electrostatic charges when in contact with differing materials.

1.4 CAPILLARY FORCES

Water readily forms hydrogen bonds with other polar materials and therefore it tends to be adsorbed to a wide variety of materials over a wide range of temperatures and humidities. However, the adsorption of water onto the majority of crystalline solids was thought to be unlikely to cause the solids to dissolve (van Campen *et al.*, 1983; Kontny *et al.*, 1987). This is because only about two layers of water molecules form on solid surfaces, and this apparently is not sufficient for dissolution to occur. Furthermore, the surface-adsorbed water exhibits a different structure to the tetrahedral structure of bulk water and thus, it will react in a different way to that expected from bulk water. However, adsorption of water molecules onto dry powders results in liquid bridges forming between the particles. Liquid capture by particles may influence interparticulate forces in at least three different ways, since (a) water uptake on a particle surface will influence the surface energy of the particle, (b) water may alter the surface conductivity and therefore the electrostatic charging of the particles and (c) water may increase interaction by capillary forces.

Capillary condensation occurs at high relative humidities, usually in excess of 60%

(Hiestand, 1966). Condensation of water vapour on the particles is mainly due to the high surface energy of fine particles. If enough water condensation occurs, a liquid bridge may form between neighbouring particles, which may greatly increase the attractive forces between particles. The interaction due to liquid bridges sometimes may be as high as several times the gravity force due to particle mass, when the surrounding atmosphere reaches the saturation pressure of water vapour.

For two smooth, spherical particles with radius r that are wetted by a liquid with a surface tension of γ and a contact angle α (Figure 1.2), the force due to a liquid bridge can be calculated as:

$$F = 2\pi\gamma r \cos\alpha \qquad (1.17)$$

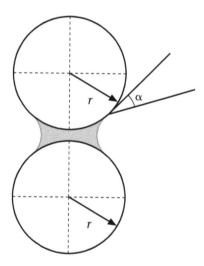

Figure 1.2: Diagrammatic representation of liquid bridging between two spherical particles.

Similarly, if water vapour condenses on the interface between a particle and a surface (Figure 1.3), the liquid bridge or meniscus thereby formed, will add to the overall adhesion force, F_c, by means of surface tension:

$$F_c = 2\pi\gamma r(\cos\alpha + \cos\beta) \qquad (1.18)$$

where α and β are the contact angles of the liquid with the surface and particle respectively (Figure 1.3).

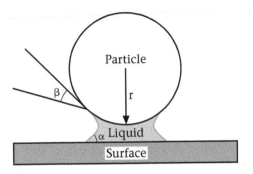

Figure 1.3: Diagrammatic representation of capillary interaction between a particle and a surface.

1.5 MECHANICAL INTERLOCKING

Pharmaceutical solids are rarely smooth since the particle surface usually abounds in asperities, i.e. 'hills and/or valleys'. When these particles come together, these asperities on the surface of different particles can interlock with each other and hence increase the particulate interactions (Figure 1.4). Further, particle contact will occur at relatively few points, which may constitute a tiny fraction of the total apparent area of the particles. The load at these points is directly proportional to the force applied and as the forces are concentrated over a small surface area, the pressure acting on these parts of the particle surface can be extremely high. The local pressure imposed on the contact area may exceed the yield value of the material, such that it will flow. The flow will result in an increased area of contact, which in turn will allow a greater distribution of the load, until it is no longer sufficient to cause the solid to yield. Such a particle deformation will

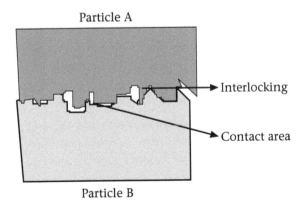

Figure 1.4: Diagrammatic representation of mechanical interlocking between two particles (A and B). As can be seen, the actual contact area between the two particles with a rough surface is only a fraction of the total surface area.

CHAPTER 1

undoubtedly increase the particulate interaction not only due to increased van der Waals forces but also by mechanical interlocking, which is more significant when the surfaces become rougher. Thus, particles having rougher surfaces will have a higher tendency to interlock mechanically than particles having smoother surfaces. If small particles are adhered to larger particles and the dimension of the asperities on the surface of the large particles are larger than the diameter of the smaller particles, the smaller particles may be entrapped in these asperities and the interaction will be much more significant than that due to intermolecular and electrostatic interactions.

1.6 FACTORS AFFECTING PARTICULATE INTERACTIONS

Given the complexity of the mechanisms of particulate interactions, it can be expected that many factors are involved in determining the overall interparticulate forces. These include the nature of the interactive particles, the state of their interaction, processing and storage conditions, to name but a few. Although it is not possible to list all the factors that may have an impact on such interactions, some of the major ones can be outlined. For example, the morphology of interactive particles (such as particle size, shape, surface texture) and their physico-chemical and mechanical properties (such as hygroscopicity, hardness, electrical properties, etc.) are among the primary factors governing powder properties such as flowability and compressibility.

1.6.1 Particle size

It is obvious that the larger an object such as a particle is, the stronger the influence it will exert on adjacent objects, and hence the stronger the interparticulate forces. If the Hamaker or Lifshitz–van der Waals constants, separation distance and particle diameters are known for a given system, then the resultant van der Waals forces can be easily calculated according to equation (1.4).

From Table 1.2, it can be seen that van der Waals forces increase with particle size, in other words, the larger the interactive particles, the stronger their interaction due to such forces. Therefore, it appears to be sensible to decrease the particulate interaction simply by means of reducing the size of the interacting particles. However, the performance of a powder composed of fine particles is often determined by the relative importance of the interparticulate forces to the gravitational forces acting upon the interacting particles. Whilst the van der Waals forces are directly proportional to the particle size (d, see equations (1.4) and (1.7)), the gravitational forces are proportional to the cube of the particle size (d^3). Therefore, gravitational forces decrease more rapidly with a decrease in particle size as compared with van der Waals interactions. Thus, in the case of large particles, van der Waals interactions may be less significant than gravitational forces. When the particle size is reduced to such a range that particulate

TABLE 1.2

Calculated van der Waals and gravitational forces for different sized particles

Van der Waals forces between two spheres (in 10^{-8} N) when $A = 1$ eV				
Particle Size (μm)	Separation distance (nm)			
	0.5	5	10	63
0.4	0.5	5	10	63
1	0.075	0.75	1.5	9.45
10	0.00075	0.0075	0.015	0.0945
Particle Density (g cm^{-3})	The gravitational force acting on a sphere (in 10^{-8} N)			
1	6×10^{-8}	6×10^{-5}	0.0004	8
10	6×10^{-7}	6×10^{-4}	0.004	80

interactions begin to become evident, then, it will start to affect the powder behaviour. Generally, gravitational forces may be dominant for particles having diameters of millimetre order whilst dispersion forces may be dominant for particles of the order of a micrometer. In fact, particulate interaction due to van der Waals forces alone is dominant over gravitational forces when small particles (<10 μm in diameter) come into contact, i.e. when the separation distance r is of the order of several nanometers. For example, for particles of 5 μm in diameter, the force of attraction at a separation distance of 10 nm is about 7.5×10^{-11} N which is more than 100 times that of the gravitational forces assuming the density of the particle to be 1 g cm^{-3}. Attraction forces become even more dominant for smaller particles and/or smaller separation distances. Because of these attraction forces, fine particles are highly sticky, cohesive and have poor flowability.

Particle adhesion to a solid substrate also begins to manifest itself when gravitational forces become negligible. Sand on a beach never causes a real adhesion problem; it can be easily shaken off from a towel or a swimming suit. Writing on a blackboard with a piece of chalk, on the other hand, is possible because the chalk produces much smaller particles than those of sand. The gravitational force acting on the chalk powder is small in comparison to the adhesive forces of these particles to the blackboard. The strong adhesive forces of the chalk particles are further enhanced by their entrapment in the crevices of the blackboard surface, a phenomenon similar to mechanical interlocking. Powder adhesion is also commonly encountered in the handling of pharmaceutical solids, leading to the adhesion of powder to the walls of containers, transport tubes, etc. The problems arising from powder adhesion is aggravated by reduction in particle size, which can be explained by reference to similar mechanisms.

1.6.2 Particle shape

Particle shape is one of the most intractable and uncontrollable factors in powder technology. Different generation methods for the same material will almost always result in the production of particles of different particle shape. It is not even uncommon for particles which have undergone similar treatments to possess different shapes. So far, all the equations generated to calculate interparticulate forces (either van der Waals, capillary or electrostatic forces) are based upon the idealised situation. The assumption is made such that particles of a predetermined shape interact under hypothetical conditions. For example, the equations outlined above all assume an interaction between perfectly spherical particles with smooth surfaces (Figure 1.5a). In reality, systems are never so ideal and some of the more realistic configurations are shown in Figure 1.5b, c and d. The interaction between these irregularly shaped particles is extremely difficult, if not impossible, to calculate. Any efforts to deduce equations to calculate the interparticulate forces within a real particulate system would prove to be unrealistic. However, with the basic principles generated from idealised situations, it is possible to assess qualitatively the particulate interaction within a powdered system.

Interparticulate forces decrease as a function of increased distance of separation between the particles and hence, any means of increasing this distance would reduce particulate interaction. This is the case when two particles having small surface asperities (i.e. those not exhibiting substantial mechanical interlocking) interact with each other as shown in Figure 1.5b. Such surface asperities are likely to limit the approach of two particles and the effective separation distance will be large, thereby limiting the van der Waals attractions almost to zero when the asperities are of the order of 1 μm (Visser, 1989). In case (c), the opposite is true. The intimate contact between these elongated particles will substantially increase the contact area and reduce the effective separation distance between the particles, both of which will increase the overall van der Waals forces. The interparticulate forces within such a closely packed powder can be expected to be stronger than all the rest of the interactions. However, elongated particles may also interact with each other in a manner shown in Figure 1.5d, where particles are

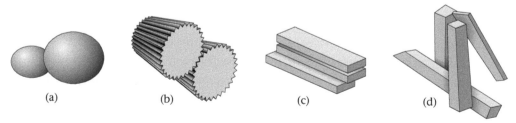

(a)	(b)	(c)	(d)

Figure 1.5: Diagrammatic representation of some typical examples of cohesion between particles. (a) Idealised situation; (b, c, d) interactions between irregularly shaped particles.

loosely packed. Such a system can be expected to have weaker particulate interactions than the other situations since the small contact area and larger separation distance in this system will reduce the attraction forces between these particles.

The packing patterns of anisometric particles, such as those with a high elongation or flatness ratio, are largely dependent upon the processing conditions that the powder has experienced. Powder handling, such as mixing, compression and vibration, may introduce enough energy to the powder such that its component particles tend to orientate themselves to the most stable conformations as shown in Figure 1.5c. Otherwise, anisometric particles tend to build up open packing of higher porosity than more isometric particles (Neumann, 1967) and hence the separation distance for anisometric particles can be expected to be larger than more isometric particles under similar conditions. Particles having a more irregular shape exhibit much higher porosities in powder beds than spherical particles both before and after tapping (Fukuoka and Kimura, 1992). Thus, the cohesive forces for anisometric particles may be smaller than those of the isometric particles unless interaction due to mechanical interlocking is implicated. However, particles of extremely irregular shape such as needles, may exhibit pronounced internal friction due to the sharp angles of the particles. Further, the cohesion of anisometric particles may be more sensitive to processing conditions than more isometric particles. Thus, the advantage of reduced interparticulate forces between anisometric particles may be greatly reduced under normal processing conditions if interlocking occurs or a higher internal friction prevails.

Due to the dependence of interaction on the relative positions of the interacting particles, reports describing the influence of particle shape on adhesion are often contradictory. Theoretically, irregular particles should exhibit higher adhesion forces than more spherical particles of a similar mass under the same conditions, if the microscopic factors are negligible and the particles seek the most stable orientation on the substrate surface. This is because of a higher contact area with, and a shorter separation distance from, the substrate surface for the irregular particles. For example, the adhesion of glass particles on a glass substrate, measured with an ultrasonic vibration technique in a dry, static-free environment, was found to decrease in the order of plates to cylinders to spheres of a given mass (Mullins *et al.*, 1992). In addition, elongated calcium carbonate particles, mixed with lactose particles (500–710 μm) at a concentration of 0.5% w/w, were found to produce more stable ordered mixes than similarly sized, more regularly shaped calcium carbonate particles, after being subjected to mechanical vibration (Wong and Pilpel, 1988). Thus, irregularity of particle shape might appear to improve the adhesion of the particle although the reverse trend has also been found. For example, the average adhesion force, f_{50}, for calcium carbonate P-70, sulphadimethoxine and silica sand particles on a glass substrate, was measured by an impact separation method (Otsuka *et al.*, 1988). The f_{50}, for all three particles, increased linearly with the shape index, Ψ, which is an indicator of sphericity of the particles. For smooth-surfaced

CHAPTER 1

particles, there was a linear relationship between log f_{50} and Ψ for particles within the size range of 25 μm $< d < 60$ μm, and a shape index $0.255 < \Psi < 0.734$, regardless of the material. This was attributed to the more angular shape of particles having lower Ψ values. The angular particles may contact the substrate with the tip or short edges of the particles and thus, the effective areas of contact may be lower as compared with those of spherical particles.

1.6.3   Surface texture

All the equations given above were based on the assumption that either two spherical particles interact with each other or that a spherical particle is adhered to a smooth, flat surface. However, real surfaces will have irregularities that change the area of contact between particles and the surface, leading to a gap between the contiguous bodies and resulting in adhesive interaction. Further, as mentioned above, in reality, solids may take a variety of shapes. The cohesive forces between irregularly shaped particles and the adhesion forces of any adherents to a surface are also a function of the interacting states between these particles or the particle and a surface. Although the adhesion of irregularly shaped particles to a rough surface is a very complicated process, the role of surface smoothness on particle adhesion may be qualitatively classified into three broad cases. The first case occurs (Figure 1.6a, c) when spherical particles adhere to a perfectly smooth surface. Only in this case may the adhesive forces be predicted using the relation outlined in equation (1.7). Such an interaction between particles and the surface may be encountered when a spherical glass particle adheres to a fused glass surface. The adhesion of some microspheres to microscope slides or even the internal walls of any container are situations where this type of adhesion occurs, although it is generally uncommon in pharmaceutical solid dosage forms. The second case (Figure 1.6b, d) occurs when the surface asperities are of small scale relative to the adherent particles. In this case the true contact area between the particles and the surface will be smaller, their separation distance will be large, and the adhesive force correspondingly less, in comparison with the values for smooth surfaces. In the third case (Figure 1.6e), the surface asperities have a larger scale than the adherent particle such that the latter is practically 'entrapped' in the former, leading to a substantial mechanical interlocking. Furthermore, the contact area can be expected to be larger and the separation distance between the particle and surface smaller. Consequently, the adhesive forces may be much higher than in both of the first two cases above. In reality, of course, when particles adhere to a plane surface they may encounter a variety of forms of surface roughness. For example, with pharmaceutical aerosol powders, fine drug particles are often attached to coarse carrier particles and some of the drug particles may adhere to the smooth regions of the carrier surface whilst others may attach to the rough sites or even fall into any gaps that can accommodate them. When exposed to external forces, such as those generated by

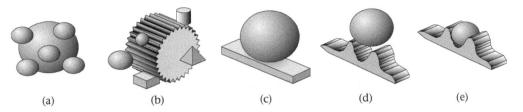

(a)   (b)   (c)   (d)   (e)

Figure 1.6: Schematic diagram showing different potential particle–particle and particle–surface interactions. (a) Spherical fine particles adhered to a spherical coarse carrier with smooth surface; (b) interaction between irregularly shaped fine particles and carrier particles with smaller scale surface asperities; (c) spherical particles adhered to a smooth planar surface; (d) interaction between a spherical particle and a surface with relatively small asperities and (e) entrapment of a fine particle within the asperities of a surface.

vibration, centrifugation, gas or liquid fluid treatment, particles adhered to smaller scale surface asperities may be easier to dislodge from the carrier particles than those attached to the smoother regions of the same particle. In contrast the 'entrapped' particles may be the most difficult to remove from the carrier. This hypothesis was partially confirmed by the fact that after the treatment of calcium carbonate particles with hydrochloric acid to produce a smoother particle surface, the average adhesion force, as measured by an impact separation method, were found to increase approximately 10-fold (Otsuka *et al.*, 1988). This was attributed to an increased surface roughness after acid treatment, which in turn reduced the contact area and increased the separation distance between particles and substrate. Interestingly, the addition of a small amount of fines (0.3–2.0% w/w) to the smooth particles with smooth surfaces before mixing with the substrate was shown to reduce the adhesion forces, producing adhesive forces similar to those of the untreated particles. The effect of fine particles was thought to be due to a similar mechanism to that brought about by the surface asperities. The presence of protuberances or fine particles on the surface may decrease the contact area and increase the separation distance between the particles and the substrate, leading to a reduction in the adhesion forces.

1.6.4 Contact area

If particulate interaction produces particle deformation, then the contact area between the interacting particles will increase. Consequently, the total van der Waals forces will be the sum of the force (F_1) between the interactive particles when no deformation occurs and the force (F_2) arising from the increased contact area due to deformation of the particles:

$$F = F_1 + F_2 = \frac{A}{12r^2}\left(\frac{d_1 d_2}{d_1 + d_2}\right) + \frac{A}{6r^3}\pi z^2 \tag{1.19}$$

where z is the radius of the contact area between the deformed particles.

Similarly, any deformation of either the particle or plane surface on contact will greatly increase the contact area which will, in turn, greatly increase the adhesive forces of the particle to the surface. After deformation, the total adhesion force will consist of two additive components, namely, the forces between the adherents before deformation at the instant of first contact and the forces acting on the contact area due to the deformation. Thus, according to Zimon (1982), the total adhesive forces (F) can be calculated from:

$$F = \frac{\sqrt{A_{11}A_{22}}}{6r^2} d_1 \left(1 + \frac{r_c}{3rd_1}\right)$$ (1.20)

where r_c is the radius of the contact area between the particle and the surface.

If elastic contact occurs when a sphere with a radius d_1 is pressed against a perfectly smooth surface with a force F_p, can be calculated by the Hertz formula (Hertz, 1896):

$$r_c = \left[0.75d_1F_p \left(\frac{1 - \mu_1^2}{E_1} + \frac{1 - \mu_2^2}{E_2}\right)\right]^{\frac{1}{3}}$$ (1.21)

where μ_1, μ_2 are the Poisson ratios for the particle and the surface, respectively, and E_1 and E_2 are the moduli of elasticity for each of the two materials.

From equation (1.20), it can be seen that the larger the contact area, the higher the adhesive force. The role of elastic or plastic deformation of particles on contact may therefore be decisive in determining the adhesion forces for the following reasons. First, easily deformable particles will exhibit higher adhesion forces than harder particles under similar conditions: this is exemplified by comparing the case of diamond and wax. The extremely hard diamond particles are not adhesive although they may possess high surface energy. On the other hand, wax, which has, predictably, a much lower surface energy than diamond, is readily adhesive, due to its high deformability on contact with another surface. The hardness of the substrate surface is as important as that of the adhered particle. For example, the adhesion of gold spheres with a diameter of 6–7 μm to a polyamide plate was found to be greater than the adhesion to a gold plate or a smooth quartz plate (for detachment of 50% of the particles, the respective forces required were 2.5×10^{-3}, 9×10^{-4} and 5×10^{-4} mN, respectively). This difference in adhesion was mainly attributed to the difference in contact area after the deposition of the gold spheres to surfaces of different hardness (Zimon, 1982).

Whether a particle will undergo elastic or plastic deformation is largely dependent upon the material properties of the particle and the extent and duration of the external forces (Rimai and Busnaina, 1995). As mentioned above, both the size and shape of the contact region between a particle and a surface are dependent upon the elastic properties of the materials (Hertz, 1896) but in addition, the surface energy of the interacting systems may also play a role in determining the particle deformation (Johnson *et al.*, 1971). For most particles, the contact area can be divided into two coaxial regions of stress. The inner zone, being subjected to greater stresses, is likely to undergo plastic deformation whereas the outer zone, experiencing a lower stress, is predicted to deform elastically (Krishnan *et al.*, 1994).

The deformation of particles on contact is affected not only by the applied pressure and particle hardness but also by the duration of the contact between the adherents. Deformation occurs as the time of adhesion is increased, particularly for polymer-like materials. The longer the time that load is applied to deformable particles, the stronger the adhesion will be. In the absence of an applied force, the increase in the adhesive interaction between a particle and a substrate surface with increasing time of contact is commonly termed 'ageing'. 'Ageing' may also be an important factor in the generation of adhesive forces and eventually, this phenomenon may affect the removal efficiency of particles from a surface. For example, Krishnan *et al.* (1994) deposited submicrometer (0.5 µm) diameter polystyrene latex particles on a silicone wafer and measured the contact radius and particle removal as a function of time (Figure 1.7).

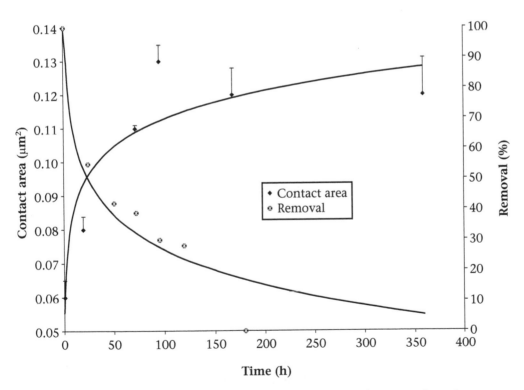

Figure 1.7: Contact area and removal efficiency of submicrometer polystyrene spheres from a silicone substrate as a function of time (adapted from Krishnan *et al.*, 1994).

CHAPTER 1

It can be seen that, as the contact area between polystyrene particles and the silicone substrate increased with time, the removal efficiency showed an inverse dependence as contact time was extended. This decreased removal efficiency was attributed to higher contact areas and increased adhesion due to the prolonged period of contact.

1.6.5 Surface energy

A solid, by definition, is a material that is rigid and resistant to stress. Similar to liquid surfaces, solid surfaces also possess a net imbalance of surface forces and therefore have a surface energy. However the surface molecules are reasonably free to move in a liquid, and consequently, there is a consistent surface energy over the entire surface. However, molecules of a solid are held together much more rigidly and are thus less free to move. Therefore, the surface energy is not evenly distributed over the solid surface. Some regions may possess a higher surface energy, for example at the edges and surface asperities on a particle, than other regions, such as plane surfaces.

The immobility, at least on a large scale, of the surface atoms of a refractory solid has the consequence that the energy of the surface and its other physical properties depend greatly on the immediate history of the material. For example, the shape of a solid is dependent upon its previous history such as whether it has been crystallised, milled or been prepared by condensation, rather than on its surface energy. A clean cleavage surface of a crystal will have a different (and probably lower) surface energy than a ground or abraded surface of the same material, and this will in turn be different from that after heat treatment. Surface defects are another history-dependent aspect of the condition of a surface. Actual crystals are by no means perfect; they may be composed of microcrystalline domains bounded by slip planes, or possess lattice irregularities, which may involve extra or missing layers or a screw or spiral dislocation. Such imperfections and dislocations leave their mark on the surface and they may possess higher surface energy than other regions on the crystal surface.

Another major difference between liquid and solid surfaces is that for liquids, the surface tension and surface energy are numerically equal, whilst for solids the surface tension is not necessarily equal to the surface energy. The surface tension is the work required to form a unit area of a surface whilst the surface energy involves the work spent in stretching the surface. The process to form a fresh surface may be roughly divided into two steps. First, the solid or liquid is cleaved so as to expose a new surface, keeping the atoms fixed in the same positions that they occupied when in the bulk phase. Second, the atoms in the surface region are allowed to rearrange to their final equilibrium positions. In the case of a liquid, these two steps occur as one, but with solids the second step may occur only slowly because of the immobility of the surface atoms or molecules. Thus, with a solid, it may be possible to stretch or compress a surface region without changing the number of atoms in it, only their distance apart.

Qualitatively, the explanation of surface energy in terms of molecular interactions is straightforward. When a new surface is produced by a process of division at 0°K, the energy of the crystal is increased by the reversible work required to form the surface. Generally speaking, this energy must be supplied to break the 'bonds' between molecules on opposite sides of the incipient surface. The formation of the surface can be visualised as occurring in two steps. An infinite crystalline solid is first split to produce two new faces maintaining the structure of each part the same. This is followed by a relaxation of the surface region, which occurs spontaneously and results in a decrease in the surface energy. It is convenient to write the specific surface energy at 0°K (E_0^S) as:

$$E_0^S = E^{S0} + \Delta E^S \tag{1.22}$$

where E^{S0} is the value of the surface energy estimated on the assumption that the configuration is uniform throughout the crystal. The term ΔE^S is the correction arising from the distortion which occurs in the surface region.

An alternative view is to consider the crystal which already has a surface present. The particles in the surface region are less tightly bound than those in the interior, since in general they have fewer neighbours. The specific surface energy is then equal to the total excess energy expressed per unit area and can be represented again by a sum of the above two terms E^{S0} and ΔE^S. As most particulate interactions in pharmaceutical solids are of the surface–surface type, their surface energies are important factors in determining the interaction forces.

The minimum work required for the separation of two surfaces and, therefore, the energy bonding them together is equal to the difference in free energy before and after separation. The work of adhesion, W_a, between two materials C and D is given by:

$$W_a = A \left(\gamma^C + \gamma^D - \gamma^{CD} \right) \tag{1.23}$$

where A is the area produced by the separation, γ^C and γ^D are the free energies per unit surface area of solids C and D in air, respectively; and γ^{CD} is the free energy of the C–D interface per unit area.

Thus, for the same substance C, the work of cohesion, W_c is given by:

$$W_c = 2A\gamma^C \tag{1.24}$$

Therefore, both cohesion or adhesion of solids are related to the surface energies of the interacting objects. Solids with a high surface energy have a high tendency to adsorb other materials onto their surfaces and form strong bonds with adhered particles (Sutton, 1976). A material shows its highest surface energy when its surface is completely clean.

■ CHAPTER 1 ■

Thus, any contamination of particle surfaces by gas, liquid or solid particles would reduce the adhesive forces between adhered particles as long as liquid and/or solid bridging is not involved. For example, when two clean iron surfaces are pressed together using a load of only a few grams, over 500 g wil be required to pull them apart. Stepwise addition of traces of oxygen reduced the cohesion and even hydrogen and helium have been shown to reduce the surface interaction (Bowden and Tabor, 1953). Another example demonstrating the interaction between solid surfaces involves molecularly flat surfaces of mica. Freshly cleaved but uncontaminated mica surfaces can be put back together with very little loss in strength. However, if the sheets are separated and coated with a monolayer of a fluorinated fatty acid and then placed together, the work required to separate them will be reduced. Such a reduction in separation forces is a reflection of the reduced cohesion forces between the mica sheets due to the contaminants (Hiestand, 1966).

1.6.6 Hygroscopicity

Particle–particle or particle–substrate interactions that are the result of capillary forces will be affected by the contact angle of the liquid on the different surfaces. According to equations (1.17) and (1.18), capillary forces have their greatest effect on adhesion or cohesion in the case of hydrophilic particles and surfaces, when the contact angle of the liquid on the particles and/or surface $\rightarrow 0°$, and have their least effect on the adhesion between hydrophobic particles and surfaces, when the contact angles $\rightarrow 90°$.

When a liquid is placed on the surface of a solid, it may either spread or form a lens, depending on the properties of both the liquid and the solid. The tendency for a liquid to spread is estimated from the magnitude of the contact angle (θ), which is defined as the angle formed (projected through the liquid) between the tangent drawn to the drop at the three-phase interface and the solid surface (Figure 1.8).

It can be seen from Figure 1.8 that the only place where a three-phase interface occurs is around the circumference of the drop. The contact angle is a consequence of a balance of the three interfacial forces, i.e. those between the solid and liquid γ_{SL}, acting to prevent spreading; those between the solid and vapour γ_{SV}, acting to aid spreading

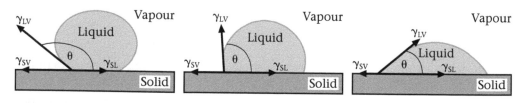

Figure 1.8: Schematic diagram of the contact angle of a liquid on a solid surface.

and those between liquid and vapour γ_{LV}, which acts along the tangent to the drop. The interfacial forces are related to contact angle by Young's equation:

$$\gamma_{LV} \cos \theta = \gamma_{SV} - \gamma_{SL} \tag{1.25}$$

A low value for the contact angle indicates good wettability, with total spreading being expressed by an angle of 0°. In contrast, a high contact angle indicates poor wettability, with the extreme being total non-wetting with a contact angle of 180°.

Since capillary forces are affected by the contact angle, hydrophilic particles are more likely to be affected by humidity than hydrophobic particles. Capillary interaction takes place over a period of time, which is mainly dependent upon the physico-chemical properties of the interactive systems and the environmental conditions, such as humidity and temperature. High humidities and temperatures will accelerate capillary condensation and an increased hygroscopicity of the interactive systems would have similar effects. Therefore, capillary interaction would be expected to increase with storage until it reaches a constant level, when the partial pressure produced by the liquid achieves equilibrium with that of the vapour in the surrounding atmosphere.

1.6.7 Relative humidity

Relative humidity can affect interparticulate forces through two opposing mechanisms. On the one hand, the environmental relative humidity increases particulate interaction due to capillary forces if a significant amount of water is condensed on the surface of the interacting systems. On the other hand, it will also increase the conductivity of both the interactive particles and the surrounding atmosphere, which will in turn accelerate the dissipation of electrostatic charges on the particles. Consequently, an increase in relative humidity will lead to a reduction in the particulate attraction due to electrostatic forces. Therefore, the effect of capillary condensation on the overall particulate interaction can be expected to follow different trends when the interacting system is exposed to environments of different relative humidities.

At low relative humidities (<50%), capillary forces usually do not contribute to the adhesive interaction. At a relative humidity between 50 and 60%, capillary forces are only beginning to occur, whilst raising the relative humidity to between 65 and 100%, leads to such forces becoming potentially prevalent over other forces and they are then likely to be the main factor in adhesive interaction (Zimon, 1982). However, excessively low humidities may increase the electrostatic charges on powders during handling and this may in turn increase adhesion (Karra and Fuerstenau, 1977). For example ambient humidities >40% were shown to decrease the degree of interaction of three sulphonamide powders with model hydroxylpropylmethylcellulose phthalate-coated glass bead carriers following blending (Kulvanich and Stewart, 1988). The reduction in adhesion was attributed to a decrease in charge due to the increase in conductivity of both the materials and the surrounding atmospheres. Substantial capillary interaction was not observed at relative humidities between 65 and 80% since the adhesion measurements were performed immediately after blending before the liquid bridges would be well established.

Capillary forces are determined by the surface tension of the liquid (usually water), the sizes of the interacting particles and the contact angle of the liquid on the surface (i.e. hydrophilicity of the particles). However, the environmental humidity is the most secure indicator of the extent of capillary forces, as condensation of water vapour on a particle surface is largely dependent upon the relative humidity to which the particles are exposed. The uptake of water molecules on particles has been divided into two stages according to the level of ambient humidity (Coelho and Harnby, 1978). At low relative humidities, the moisture associated with the particles of a powder exists as adsorbed water vapour. As the humidity is increased, the thickness of the adsorbed layer also increases until eventually condensation occurs at the contact points and liquid bridges are formed. The boundary between the two forms of water retention is determined by the liquid stability and can be derived from the BET isotherm for the adsorption of water vapour on a solid surface (Buckton, 1995). Thus, there exists a critical relative humidity above which a liquid bridge will be formed at the contact points of particles. Below this value, water vapour is adsorbed onto the surface of the particles and hence, capillary forces are expected to be negligible. Increasing the relative humidity but keeping it below the critical humidity will decrease the interparticulate forces by reducing electrostatic attraction. Charge decay will be increased as a result of the reduction in the resistivity of both the particles and the ambient atmosphere (Boland and Geldart, 1972). However, if the relative humidity is above the critical value of the particles and water vapour condensation is allowed to reach equilibrium, a further increase of relative humidity will enhance capillary forces and hence the overall interparticulate forces.

1.6.8   Electrical properties

Electrical charge arises from collision and friction amongst the particles during powder mixing and other handling processes. Because the intensive movement of the material causes the charge transfer, the electrical charge arising from this process is also called triboelectrostatic charge. With each normal collision an interfacial charge transfer of electrons occurs (Figure 1.9b). The charge remains on the points of contact when the further movement of the powder bed occurs (Figure 1.9c). The separating particles may collide again and the subsequent collisions of the charged particles are extremely unlikely to find exactly the same points on the surface for the new interparticulate contact. Thus nearly every collision causes a new interparticulate charge transfer (Figure 1.9d). If the rate of charge distribution on the particle surface is negligible as compared to the fre-

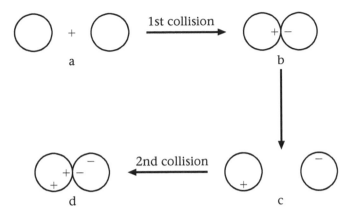

Figure 1.9: Schematic diagram showing the creation of triboelectrostatic charge. (a) Uncharged particles; (b) electron transfer between particles; (c) charged particles; and (d) electron transfer on the second collision.

quency of collision, then new charge will be continuously created until the charge density rises to such an extent that further collisions will occur on previously charged spots. As mentioned above, the charge sign of each individual group of particles will depend largely upon their relative electron-donor or -acceptor properties and work functions. Furthermore, in the case of particles of the same material, electron transfer from large particles to small particles is favoured rather than the other way around, since the former have lower work functions (Bailey, 1984).

Water vapour condensation on a particle surface will improve the distribution of triboelectrostatic charges. The thicker the water layer, the higher the conductivity of the particle and the more easily this charge distribution occurs. Crystallinity also has an effect on the charge distribution. Since molecules of an amorphous material have higher mobility than those in a crystal lattice, charge distribution often occurs faster on amorphous particles than on crystalline particles. Electrical charges may distribute relatively homogenously on amorphous spherical particles (Figure 1.10a) whereas in the case of a crystalline particle with pointed edges, most charges may concentrate on tips (Figure 1.10b). After violent collisions, the total charge on the crystals with sharp edges, especially in the region of edges or tips, may rise to extremely high values.

The contribution of electrostatic forces to the overall particulate forces is usually thought to be less important than that of the van der Waals forces. However, electrostatic forces will become more significant and may even dominate when particle size is reduced. For example, the adhesive forces of some model drug particles, namely, sulphapyridine (27.2 μm), sulphamerazine (17.7 μm) and succinylsulphathiazole (23.4 μm) to hydroxypropylmethylcellulose-coated glass beads were found to increase linearly (correlation coefficient >0.90) with the square of the average charge to mass ratio

CHAPTER 1

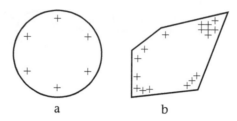

Figure 1.10: Schematic diagram showing charge distribution on different particles. (a) Spherical (amorphous or water-condensed) particle and (b) irregular crystalline particle.

(Kulvanich and Stewart, 1988). In all cases a decrease in the extent of interaction was observed during storage over 23 days due to charge decay in the systems.

In conclusion, the predominant interparticulate forces between the particles of a powder are the van der Waals forces of attraction. Electrostatic forces are only important at low environmental humidities whereas capillary forces become manifest at high humidities. All these forces will only be noticeable for particles with dimensions in the order of a few micrometers in diameter or smaller. Although the interparticulate forces can be qualitatively estimated using a mathematical model, it is practically impossible to predict the actual force of particulate interaction on a quantitative basis since many factors are involved in this process. Thus, great caution must be exercised when attempting to extrapolate findings derived from hypothetical models to real systems encountered in pharmaceutical and other processes.

REFERENCES

BAILEY, A.G. (1984) Electrostatic phenomena during powder handling. *Powder Technol.* **37**, 71–85.

BOLAND, D. and GELDART, D. (1972) Electrostatic charging in gas fluidised beds. *Powder Technol.* **5**, 289–297.

BOWDEN, F.P. and TABOR, D. (1953) In: GOMER, R. and SMITH, C.S. (eds), *Structure and Properties of Solid Surfaces*. Chicago, University of Chicago Press, 213.

BRITTAIN, H.G. (1995) Overview of physical characterization methodology. In: BRITTAIN, H.G. (ed.), *Physical Characterization of Pharmaceutical Solids*. New York, Marcel Dekker, 1–36.

BRITTAIN, H.G., BOGDANOWICH, S.J., BUGAY, D.E., DEVINCENTIS, J., LEWEN, G. and NEWMAN, A.W. (1991) Physical charaterization of pharmaceutical solids. *Pharmaceut. Res.* **8**, 963–973.

BUCKTON, G. (1995) *Interfacial Phenomena in Drug Delivery and Targeting*. Chichester, Harwood Academic Publishers.

COELHO M.C. and HARNBY, N. (1978) The effect of humidity on the form of water retention in a powder. *Powder Technol.* **20**, 197–200.

DZYALOSHINSKII, I.E., LIFSHITZ, E.M. and PITAEVSKII, L.P. (1960) General theory of the van der Waals forces. *Sov. Physik., JETP* **10**, 161.

FUKUOKA, E. and KIMURA, S. (1992) Cohesion of particulate solids, VIII. Influence of particle shape on compression by tapping. *Chem. Pharm. Bull.* **40**, 2805–2809.

HAMAKER, H.C. (1937) London–van der Waals forces attraction between spherical particles. *Physica (Utrecht)* **4**, 1058–1072.

HERTZ, H. (1896) *Miscellaneous Papers.* London, Macmillan.

HIESTAND, E.N. (1966) Powders: particle–particle interactions. *J. Pharm. Sci.* **55**, 1325–1344.

JOHNSON, K.L., KENDALL, K. and ROBERTS, A.D. (1971) Surface energy and the contact of elastic solids. *P. Roy. Soc. A* **324**, 301–313.

KARRA, V.K. and FUERSTENAU, D.W. (1977) The effect of humidity on the trace mixing kinetics in fine powders. *Powder Technol.* **16**, 97–105.

KONTNY, M.J., GRANDOLFI, G.P. and ZOGRAFI, G. (1987) Water vapour sorption of water soluble substances: Studies of crystalline solids below their critical relative humidities. *Pharmaceut. Res.* **4**, 104–112.

KRISHNAN, S., BUSNAINA, A.A., RIMAI, D.S. and DEMEJO, L.P. (1994) The adhesion-induced deformation and the removal of submicrometer particles. *J. Adhes. Sci. Technol.* **8**, 1357–1370.

KULVANICH, P. and STEWART, P.J. (1988) Influence of relative humidity on the adhesive properties of a model interactive system. *J. Pharm. Pharmacol.* **40**, 453–458.

LANGBEIN, D. (1969) Van der Waals attraction between macroscopic bodies. *J. Adhesion* **1**, 237–245.

LIFSHITZ, E.M. (1956) Theory of molecular attraction forces between solid bodies. *Sov. Physik* **63**, 245–279.

LONDON, F. (1930) Theory and systematics of molecular forces. *Z. Physik* **63**, 245–279.

MULLER, V.M., YUSHCHENKO, V.S. and DERJAGUIN, B.V. (1980) On the influence of molecular forces on the deformation of an elastic sphere and its sticking to a rigid plane. *J. Colloid. Interf. Sci.* **77**, 91–101.

MULLINS, M.E., MICHAELS, L.P., MENON, V., LOCKE, B. and RANADE, M.B. (1992) Effect of geometry on particle adhesion. *Aerosol. Sci. Tech.* **17**, 105–118.

NEUMANN, B.S. (1967) The flow properties of powders. In: BEAN, H.S., BECKETT, A.H. and CARLESS, J.E. (eds), *Advances in Pharmaceutical Sciences*, Vol. 2. London, Academic Press, 181–221.

OTSUKA, A., IIDA, K., DANJO, K. and SUNADA, H. (1988) Measurement of the adhesive force between particles of powdered materials and a glass substrate by means of the impact separation method. III. Effects of particle shape and surface asperity. *Chem. Pharm. Bull.* **36**, 741–749.

RIMAI, D.S. and BUSNAINA, A.A. (1995) The adhesion and removal of particles from surfaces. *Particul. Sci. Technol.* **13**, 249–270.

SUTTON, H.M. (1976) Flow properties of powders and the role of surface character. In: PARFITT, G.D. and SING, K.S.W. (eds), *Characterization of Powder Surfaces with Special Reference to Pigments and Fillers*. London, Academic Press, 122.

VAN CAMPEN, L., AMIDON, G.L. and ZOGRAFI, G. (1983) Moisture sorption kinetics for water-soluble substances. 1. Theoretical considerations of heat transport control. *J. Pharm. Sci.* **72**, 1381–1388.

VAN DER WAALS, J.D. (1873) Over de continuiteit van den gas- en vloeistoftoestand. PhD Thesis, University of Leyden.

VISSER, J. (1995) Particle adhesion and removal: a review. *Particul. Sci. Technol.* **13**, 169–196.

VISSER, J. (1989) Van der Waals and other cohesive forces affecting powder fluidization. *Powder Technol.* **58**, 1–10.

WONG, L.W. and PILPEL, N. (1988) The effect of the shape of fine particles on the formation of ordered mixtures. *J. Pharm. Pharmacol.* **40**, 567–568.

ZIMON, A.D. (1982) Adhesion, molecular interaction and surface roughness. In: ZIMON, A.D. (ed.), *Adhesion of Dust and Powder*, 2nd edn. London, Consultant Bureau, 46–47.

Interparticulate Forces in Pharmaceutical Powders

Contents

2.1 INTRODUCTION

One of the main reasons for the continuing popularity of solid dosage forms is the fact that they can be introduced into the body by a number of routes (Byrn, 1982). The production of the variety of such dosage forms almost always involves the handling of powdered solids (Lachman *et al.*, 1986) since the drugs and excipients will exist in this form at the beginning, throughout and, often, at the end of the manufacturing process. Not surprisingly, the physicochemical properties of powders have attracted considerable attention but this usually addressed a collection of discrete particles with dimensions less than 100 μm.

Pharmaceutical powders are composed of either fine drug particles (those <200 μm) or mixtures of drug particles together with excipient(s). Due to the intrinsic nature of small particles, interparticulate forces within a pharmaceutical powder are often one of the primary determinants of its behaviour, influencing such properties as flow, mixing and agglomeration. It is essential to understand the various factors involved in particulate interactions between pharmaceutical solids in order to achieve the optimum powder performance.

As mentioned in chapter 1, particulate interactions can be broadly classified as either cohesive or adhesive. For instance, comminution of a drug powder may result in the generation of cohesive forces between drug particles whilst, in contrast, the processing of mixtures of drug with excipients may be influenced by both types of forces. Particulate interactions may be due to a number of mechanisms, namely van der Waals, electrostatic and capillary forces, as well as mechanical interlocking. The relative contribution of each individual mechanism to the overall interparticulate force is a function of the physico-chemical and morphological properties of the interacting particles, the powder processing conditions and the environmental relative humidities. For example, at low relative humidities, van der Waals forces may prevail over other forces whilst at high relative humidities, capillary forces may become predominant if significant water vapour condensation occurs on the particle surface. If charge is generated on fine particles, especially when the particles are subject to violent movement, then electrostatic forces may be the most important factor in determining the overall interparticulate forces. However, if irregularly shaped particles with rough surfaces are brought together to such an extent that mechanical interlocking is involved in the particulate interaction, as in the case of tabletting, then the forces due to solid bridging may be several orders of magnitude higher than other forces.

Figure 2.1 summarises some of the important factors that may determine interparticulate forces, and which may in turn affect the performance of pharmaceutical powders. These may include the physico-chemical properties of the interacting powder (such as the mechanical and electrical properties and hygroscopicity); particle morphology (such as particle size, shape and surface texture); crystal forms (such as crystallinity and the occurrence of polymorphism); and external conditions (such as the environmental relative humidity and processing conditions).

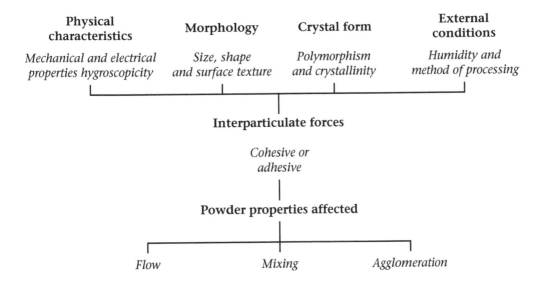

Figure 2.1: Flow chart showing some factors that may affect the interparticulate forces in pharmaceutical solids.

CHAPTER 2

2.2 MECHANICAL PROPERTIES

Pharmaceutical solids may be described as elastic, plastic, viscoelastic, hard, tough or brittle. All of which will have a profound effect on solid processing. For example, tabletting properties are influenced by elastic and plastic deformations as well as the viscoelastic properties of the powder. Powder flow may also be determined by these mechanical properties since soft materials usually have poorer flowability than harder materials of similar particle size.

Elastic deformation refers to a change in shape that is completely reversible, that is to say, the particle will return to its original shape upon release of the applied stress. Elastic deformation usually occurs during initial stages of stress application. During elastic deformation, the deformation stress, σ_d, is described by Hook's law:

$$\sigma_d = \epsilon E \tag{2.1}$$

where E is the Young's Modulus of Elasticity, and ϵ is the deformation strain.

If an isotropic body is subjected to a simple tensile stress in a specific direction, it will elongate in that direction whilst contracting in the two lateral directions, and its relative elongation is directly proportional to the stress. The ratio of the stress to the relative elongation (strain) is termed Young's Modulus of Elasticity, which is a fundamental property of a material directly related to its interatomic or intermolecular binding energy, and is a measure of its rigidity. For materials with a high Young's Modulus of Elasticity, relatively high stresses induce only small changes in length (strain). Materials with low elastic moduli are normally described as 'elastic' materials whilst those with high elastic moduli are regarded as 'stiff' materials.

The Young's modulus of a material can be determined by many techniques, such as flexure or indentation testing on either crystals or compacts (Rowe and Roberts, 1995). In flexure testing, particles are first compressed at varying pressures to prepare rectangular beams of varying porosity. Then, a beam of small thickness and width in comparison with its length is subjected to transverse loads and its central deflection caused by bending is measured. Young's modulus for a specific porosity can be calculated from the applied load and the deflection of the mid-point of the beam. If Young's modulus is plotted against the porosity then the true modulus of the material can be obtained by extrapolation to zero porosity. Young's modulus can vary over two orders of magnitude for hard rigid materials with very high moduli (e.g. metals) to soft elastic materials with very low moduli (e.g. polymers) (Table 2.1). For example, tabletting excipients can be placed in the following increasing order of rigidity: starch < microcrystalline celluloses < sugars < inorganic fillers. Between and within groups variation can occur depending upon chemical structure, method of preparation and particle size. For example, different polymorphic forms of lactose exhibit different rigidity in an increasing order of spray dried < β-lactose < α-lactose monohydrate. Interestingly, virtually all the drugs tested to date exhibit lower moduli than the cellulosic materials. This might be explained by the fact that many organic solids including drugs form glasses which exhibit anomalous endotherms that resemble glass transitions and can therefore be regarded as possessing a certain amount of mobility.

Plastic deformation refers to the permanent change in shape of a specimen due to an applied stress and its onset is seen as curvature in the stress–strain curve. Plastic deformation is important because it allows pharmaceutical excipients and drugs to establish realistically large areas of contact during compaction that remain on decompression. This property is crucial for the production of tablets. When plastic deformation occurs, the deformation stress, σ_d, is equal to the yield stress of the material.

The hardness of a material is usually measured by the indentation method in which a hard indentor (e.g. diamond, sapphire, quartz or hardened steel) of a specified geometry is pressed into the surface of the material (Rowe and Roberts, 1995). The hardness is essentially the load divided by the projected area of the indentation to give a measure of the contact pressure. The sample can again be either a compact or a single crystal of the material to be tested. Although compacts have been widely employed to measure hardness, their use is associated with several artefacts since compression may alter some of

TABLE 2.1

Young's modulus for different pharmaceutical excipients and drugs determined using beam bending methods

Materials	Young's modulus (GPa)	Particle size (μm)	References
Celluloses			Bassam et al. (1990)
Avicel PH1101	9.2	50	
Avicel PH102	8.7	90	
Avicel PH105	9.4	20	
Emcocel	7.1	56	
Emcocel (90M)	8.9	90	
Unimac (MG100)	8.0	38	
Unimac (MG200)	7.3	103	
Sugars			
α-Lactose monohydrate	24.1	20	Roberts et al. (1991)
β-Lactose	17.9	149	Bassam et al. (1990)
Spray dried lactose	11.4	125	Bassam et al. (1990)
Dipac Sugar	13.4	258	Bassam et al. (1990)
Mannitol	12.2	88	Bassam et al. (1990)
Polysaccharides			
Starch 1500	6.1	–	Roberts and Rowe (1987)
Maize starch	3.7	16	Bassam et al. (1990)
Inorganics			
Emcompress	181.5	–	Roberts and Rowe (1987)
Calcium carbonate	88.3	8	Bassam et al. (1990)
Calcium carbonate	47.8	10	Bassam et al. (1990)
			Roberts et al. (1991)
Polymers			
PVC	4.1	–	
Stearic acid	3.8	62	
Drugs			Roberts et al. (1991)
Theophylline (anhydrous)	12.9	31	
Caffeine (anhydrous)	8.7	38	
Sulphadiazine	7.7	9	
Aspirin	7.5	32	
Ibuprofen	5.0	47	
Phenylbutazone	3.3	50	
Testosterone propionate	3.2	85	

the intrinsic properties of the materials concerned. Further, similar to Young's modulus, the hardness of a material measured by indentation is also affected by the compact porosity. Therefore, the use of a single crystal to measure hardness may be the preferred means of measuring the true hardness of the material since values measured on compacts are usually slightly different. Ichikawa et al. (1988) attributed such differences in hardness to the mechanism of deformation during compaction.

As with the modulus of elasticity, values for hardness of pharmaceutical excipients vary over two orders of magnitude from the very hard materials (e.g. Emcompress) to very soft waxes. Crystalline sugars usually exhibit the highest hardness values (Table 2.2).

TABLE 2.2

Indentation hardness measured on compacts and single crystals

Material	Indentation hardness (MPa)	
	Compacts	Single crystals
Sugars		
α-Lactose monohydrate	534[a]	523[b]
β-Lactose	251[c]	
Sucrose	1046–1723[c]	
Drugs		
Paracetamol	265[a]	342[b]
Caffeine (anhydrous)	290[a]	
Caffeine (granulate)	288[a]	
Hexamine	232[c]	42[b]
Phenacetin	213[c]	172[b]
Others		
Emcompress	752[a]	
Avicel PH102	168[a]	
Sodium stearate	37[d]	
PEG 4000	36[d]	
Magnesium stearate	22[d]	
Sodium lauryl sulphate	10[d]	
Urea		93[b]

[a]Jetzer *et al.* (1983).
[b]Ichikawa *et al.* (1988).
[c]Leuenberger (1982).
[d]Leuenberger and Rohera (1985).

The relationship between elasticity and hardness of a material can be expressed by the following semi-empirical equation derived by Marsh (1964):

$$\frac{H}{\sigma_y} = 0.07 + 0.61 \ln\left(\frac{E}{\sigma_y}\right) \tag{2.2}$$

where H, E and σ_y are the indentation hardness, Young's modulus and yield stress, respectively.

Thus, the higher the Young's modulus (i.e. the less elastic the material) the higher the hardness value of the material. Highly elastic materials, such as polymers, have values of E/σ_y between 10 and 15, and will exhibit values of H/σ_y between 1.5 and 1.0. Whereas if E/σ_y is between 25 and 30, then H/σ_y is between 2.0 and 2.2 and materials with such values are brittle, typical of glasses. If $E/\sigma_y > 150$ then $H/\sigma_y > 3.0$, and this represents a rigid-plastic material typical of metals (Rowe and Roberts, 1995).

The mechanical properties of materials may be one of the most important factors in determining particulate interactions. Interparticulate forces due to van der Waals forces are a function of contact area between the interacting particles (see equation (2.3)):

$$F = \frac{A}{12r^2}\left(\frac{d_1 d_2}{d_1 + d_2}\right) + \frac{A}{6r^3 \pi} A_c \tag{2.3}$$

where F is the van der Waals force, A_c the contact area between the deformed particles, A is the Hamaker constant, d_1 and d_2 are the diameter of the two interacting particles and r is the separation distance between the particles.

For spherical particles that undergo elastic deformation, the area of contact A_c may be estimated as (Morgan, 1961):

$$A_c = \pi \left[0.75 \frac{d_1 d_2}{d_1 + d_2} F_p \left(\frac{1 - \mu_1^2}{E_1} + \frac{1 - \mu_2^2}{E_2} \right) \right]^{\frac{2}{3}} \tag{2.4}$$

where μ_1, μ_2 are the Poisson ratios for two interacting particles, E_1, E_2 are their moduli of elasticity, d_1 and d_2 are their diameters and F_p is the interparticulate force between the particles.

Thus, the higher the moduli of elasticity, i.e. the harder the material, the smaller the contact area between two interacting particles. The smaller contact area between two harder particles will in turn lead to weaker interparticulate forces. It follows that, particles of harder materials will exhibit smaller interparticulate forces than those of softer materials and this is one of the main reasons that the former usually have better flow properties than softer particles of similar size and density. From Tables 2.1 and 2.2, it can be seen that calcium carbonate is harder than many other pharmaceutical excipients such as PEG 4000. Therefore, particles of the former material may exhibit lower interparticulate forces than those between particles of the latter material.

By way of proof, the geometric median adhesion force measured using a centrifugation method was found to decrease with the hardness of materials (Lam and Newton, 1991).

Figure 2.2 shows that the median adhesion forces of powders increases proportionally as a function of compression force. The slope of each line expresses the change of the adhesion force per unit increase in compression force, the order being PEG 4000 > starch 1500 > spray-dried lactose > heavy calcium carbonate. This difference in the sensitivity to a change in compression forces was attributed to the different hardnesses of these materials (Lam and Newton, 1993). PEG 4000 was the softest, with a yield pressure of 55 MPa (Al-Angari et al., 1985), leading to the largest adhesion forces. Heavy calcium carbonate on the other hand was the hardest with a yield pressure of 486 MPa (York, 1978) and hence, it was the least deformable of the materials and also exhibited the smallest adhesion forces.

2.3 ELECTRICAL PROPERTIES

Most pharmaceutical powders are organic crystals which behave as insulators under ambient conditions and electrical charges may be conferred during nearly every powder handling process. However, triboelectrification in pharmaceutical powders is a very

■ CHAPTER 2 ■

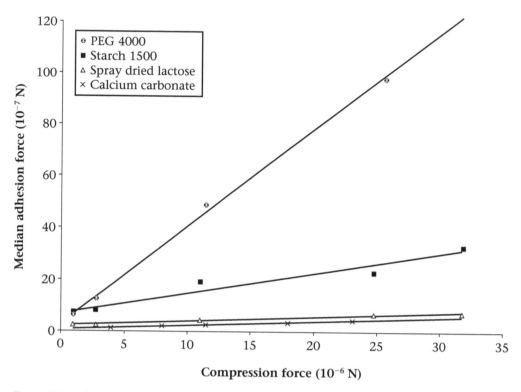

Figure 2.2: Relationship between the geometric median adhesion, measured by a centrifugation method, and the compression force (adapted from Lam and Newton, 1991).

complicated and ill-defined process although it has been shown to be influenced by many factors, such as particle size and shape, physico-chemical properties of the contacting surface, contact area and frequency, surface purity and atmospheric conditions (Bailey, 1984). The major processes that may induce triboelectrification in pharmaceutical powders involve the preparation of fine particles (e.g. mechanical micronisation or spray drying), mixing and fluidisation in a turbulent air stream. During mechanical micronisation, particle size is usually reduced by means of collisions between coarse particles so that the larger particles are broken up and fine particles are produced. The violent collisions induce high electrical charges on the micronised particles. Spray-dried particles may also be highly electrically charged since solutions are aerosolised into small droplets which are then dried in a hot air cyclone.

Electrical charges may also result from interactions between particles and containers. And depending on the nature of the latter either positive or negative charges can be produced. Most pharmaceutical excipients, including lactose, are charged electronegatively following contact with a glass surface, which usually acts as an electron-donor (Staniforth and Rees, 1982a) whereas such excipients were found to become electropositive after contact with a polyethylene surface. Both salbutamol sulphate and lactose particles became positively charged after contact with plastic containers, whereas following contact with a metal surface, lactose particles adopted an electronegative charge whilst salbutamol sulphate became positively charged. Stainless steel was found to impart a higher electronegative charge on lactose particles than brass, possibly due to the lower work function of the former (Carter *et al.*, 1992) (Figure 2.3). However, salbutamol sulphate charged electropositively when in contact with brass but electronegatively with stainless steel, but beclomethasone dipropionate particles showed the opposite charging tendency.

Some pharmaceutical powders were found to have charges at least 100 times greater after triboelectrification in an air cyclone. For example, the mean specific charge (the ratio of electrical charge to powder mass) of lactose particles (500–710 µm) decreased from $-0.63 \times 10^{-9} \, C \, g^{-1}$ after pouring off a glass surface to $-470.0 \times 10^{-9} \, C \, g^{-1}$ after fluidisation in an earthed brass cyclone (Staniforth and Rees, 1982b). Both the feeder rates of powders and the feed pressure were shown to affect powder-specific charge after fluidisation in an air cyclone (Carter *et al.*, 1992), and increasing powder feed rate decreased the particle charge, possibly by reducing the number of particle–cyclone collisions. The specific charges of powdered α-monohydrate lactose, micronised salbutamol sulphate and beclomethasone dipropionate particles were found to increase linearly with the pressures employed to fluidise these particles in a steel or brass cyclone (Carter *et al.*, 1992). Smaller lactose particles (45–90 µm) showed markedly higher specific charges than those of larger particles (355–500 µm) under similar fluidisation conditions. This latter phenomenon was attributed to the greater particle number density in the case of the smaller particles, leading to an overall increase in the probability of particle collision.

2.4 WATER VAPOUR CONDENSATION

The physico-chemical properties of pharmaceutical solids are often altered by the adsorption and/or absorption of water vapour. The sensitivity of a powder to environmental humidity is, in turn, determined by its hygroscopicity, which is defined as the potential for and the rate of moisture uptake (Carstensen, 1993). In general, the higher the relative humidity, the more a solid will sorb water vapour at a higher uptake rate.

For non-hydrating, water-soluble, crystalline solids, the water is mainly adsorbed onto the particle surface at low relative humidity (Figure 2.4a). As the relative humidity is increased, further water vapour uptake will result in the formation of multilayer condensation of water vapour on the particle surface (Figure 2.4b). Then, the solid will begin to dissolve in the sorbed film of water, until a saturated solution of the solute is formed. Dissolution of the material in the adsorbed water layer will depress the water vapour pressure over the sorbed film until it reaches the value of the vapour pressure of

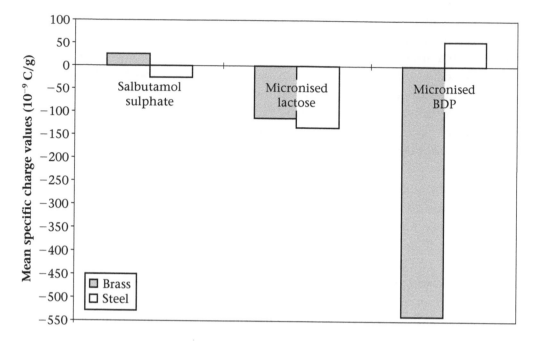

Figure 2.3: The charge to mass ratios of salbutamol sulphate, micronised beclomethasone diproprionate and lactose after fluidisation at a feeder pressure of 8.68 p.s.i. (adapted from Carter *et al.*, 1992).

a saturated solution of the specific material, which is defined as the critical relative humidity, RH_0. If the environmental relative humidity is greater than the RH_0, the condensation of water vapour on the particle surface is thermodynamically favourable. The spontaneous condensation of water vapour on the particle surface will dilute the saturated water layer, allowing more solid to dissolve, which will maintain the pressure gradient between the surrounding atmosphere and the particle surface. This process will continue until all the solid has dissolved and such a phenomenon is called deliquescence (Figure 2.4c).

The adsorption of water vapour onto the surface of hydrated water-soluble particles follows a similar trend prior to hydration when the environmental relative humidity is less than the RH_0 of the material. When the RH exceeds RH_0, condensation of water vapour on the particle surface may not result in the formation of a multilayer and instead, the sorbed water may hydrate the hitherto unhydrated particle. Since water vapour will also penetrate into the crystal lattice of the particles, relatively large quantities of water vapour may be sorbed by hydrated water-soluble particles. Thus, for these materials, water vapour may penetrate or diffuse into the particles (Sutton, 1976) before any significant liquid bridge can be formed on the surface of the particles. For example, disodium cromoglycate, a hygroscopic antiasthmatic drug, was found to accommodate

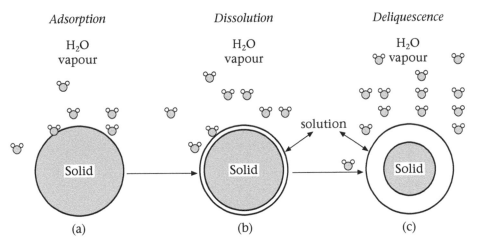

Figure 2.4: Water vapour condensation on the surface of a water-soluble particle at different relative humidities (RH). (a) RH < RH_0; (b) RH = RH_0; (c) RH > RH_0.

most of the sorbed moisture within its particles and only a small amount of water was adsorbed onto the particle surface (Chan and Pilpel, 1983). Therefore, the formation of liquid films on the surfaces of highly hydrophilic drug particles was reported to be unlikely under normal ambient conditions and, consequently, liquid bridging was assumed to be less significant for hydrophilic drugs than for hydrophobic drugs (Sutton, 1976) under normal conditions. However, when the relative humidity exceeds the RH_0 of the material, then, hydrophilic particles will sorb more water vapour than hydrophobic particles and consequently, more water will condense on the surface of the former particles than on that of the latter.

The water sorption profiles of amorphous solids such as cellulose, starches, PVP, gelatin and some lyophilised proteins, are different from those of the crystalline solids. These substances may sorb significant quantities of moisture (e.g. 25–50%) even at low environmental relative humidities. Unlike water-soluble crystals, these materials may not dissolve in the sorbed water but undergo significant morphological changes such as swelling, after considerable water sorption has occurred. A typical water sorption isotherm is shown in Figure 2.5, and it can be seen that these substances sorb significant amounts of water over the entire range of relative humidities. The hysteresis loop between the sorption and desorption isotherms indicates that the amount of water in the solid is greater for the desorption isotherm than the sorption isotherm for a given relative humidity. This is generally attributed to either kinetic effects or to a change in the polymer-chain conformation caused by the sorbed water (Zografi and Kontny, 1986). Therefore, in such cases, sorbed water penetrates throughout the amorphous solid and there is a specific water–solid interaction (Kontny and Zografi, 1985).

The water sorption profiles of hydrophobic, water-insoluble particles are much

CHAPTER 2

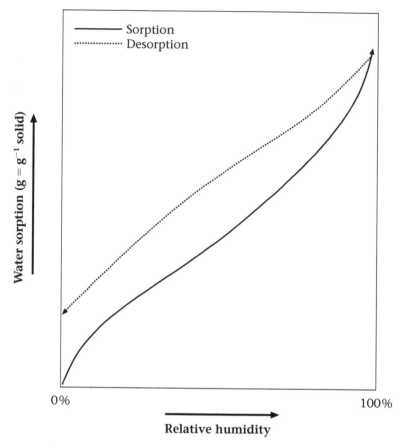

Figure 2.5: Diagram showing the water sorption and desorption profiles for a typical amorphous macromolecular material.

simpler. Since the penetration of sorbed water into the particles and dissolution of the solute in the sorbed water is negligibly small, water vapour condensation on the particle surface will reach an equilibrium where the sorbed water layer produces a partial vapour pressure that equals the environmental partial pressure. Therefore, water vapour sorption for these substances is much smaller when compared to more hydrophilic materials.

The sorption of water vapour by a powdered material will increase the interparticulate forces between the component particles through both liquid bridging and fusion among adjacent particles as a result of the wetted and softened surfaces. For example, lactose, dextrose and sucrose particles were found to be free flowing under relative humidities less than about 60% (Shotton and Harb, 1966) and particle cohesiveness was found to increase with environmental relative humidity. Both dextrose and sucrose solids eventually became solutions when stored over 90% relative humidity. However, if the particles are highly hydrophobic then the adsorbed water will not penetrate but remain on the particle surface. The adsorbed moisture may then rapidly form a thin film on the particle surface and thus increase the interparticulate forces through capillary forces. The effects of relative humidities on the cohesiveness of particles of different physico-chemical properties were systematically investigated by Fukuoka *et al.* (1983). The cohesiveness of some hydrophilic, water-soluble powders (sulpyrine, sodium bicarbonate, methylcellulose, lactose and monosodium glutamate) as a function of water uptake were found to change very slightly up to a relative humidity of 80% but an abrupt increase in both cohesiveness and water uptake was observed above this value. Hydrophilic but water-insoluble powders (α-alumina, microcrystalline cellulose, silica gel, glass powder and glass beads) showed only a very small change in cohesiveness and water uptake under low humidities (<60%) but they exhibited different trends when exposed to higher humidities. For example, α-alumina showed a drastic increase in cohesiveness but practically no change in water uptake when the relative humidity was over 60% whereas the cohesiveness of silica gel showed a slight maximum between relative humidities of 70 and 80% although water uptake increased progressively as a function of relative humidity. This phenomenon may be due to different retention states of moisture by the particles. Moisture may be condensed exclusively on the surface of alumina particles, leading to a maximum increase in capillary forces with increasing relative humidity. However, due to the porous structure of silica gel particles, a large amount of water vapour may be absorbed in such micro-pores and hence any condensed water vapour is unable to form liquid bridges between particles, resulting in relatively low capillary forces even under high relative humidities. Slightly water-soluble and hydrophobic powders (sulphamine, phenacetin, sulfisoxazole, aspirin and benzoic acid) showed a moderate increase in cohesiveness at about 40% relative humidity and a more drastic increase over 70%, although water vapour uptake by these particles remained practically constant up to this value. These results are indicative of a higher sensitivity of the cohesiveness of hydrophobic particles compared to hydrophilic particles under lower relative humidities since intact liquid bridges (providing the contact angle, α, is small) are more likely to form on the surface.

During handling, if powders are exposed to high humidities for a sufficient period of time to allow a substantial amount of water vapour to be adsorbed and they are then placed in lower humidity conditions or subjected to any drying process, liquid bridging may ultimately result in much stronger solid bridging due to the precipitation or crystallisation of dissolved molecules at the interfaces. This is particularly true for hydrophilic particles whereas any contamination with soluble materials at the interfaces between hydrophobic particles may still lead to the similar problems.

2.5 PARTICLE MORPHOLOGY

Although both van der Waals and capillary forces increase with the size of the interacting particles, interparticulate forces usually do not become apparent until the particle size diminishes to a certain extent. This is because powder behaviour is more dependent on the relative importance of interparticulate forces compared to the gravitational forces acting on the particles. Pharmaceutical solids usually range from 1 to 1000 μm in size. An important dividing line between the properties of particles of various sizes lies at about 10 μm in diameter. The van der Waals attractive force between two polystyrene particles of 10 μm diameter is of the order of 10^{-15} to 10^{-8} N, depending on the distance between them. Compared to this, the gravitational force on the same particles is about 10^{-12} N. Thus, for particles of this size, the van der Waals attractive forces are generally greater than the force of gravity, and consequently the material is cohesive. When the particles are larger, the gravitational force, which increases more quickly as a function of particle size than the interparticulate force, becomes dominant over the van der Waals forces. In this case, the material is not cohesive unless other attractive forces, such as capillary forces, are significantly involved.

The cohesiveness of fine particles is more sensitive to humidity than relatively coarser particles (Fukuoka *et al.*, 1983), despite the capillary force being proportional to the particle radii of two interactive particles. There are a number of reasons for this. First, fine particles have higher surface energy than coarse particles of the same material and hence, the former are more likely to adsorb water vapour than the latter. Second, fine particles have a higher specific surface area than coarse particles and therefore, they have higher water uptake capacity than coarser particles. Finally, intact liquid films are more likely to form on the surface of smaller particles than on that of larger particles.

As discussed in the previous chapter, the effects of surface smoothness on interparticulate forces follow many complicated mechanisms, depending on the relative scale of the surface roughness to the diameter of interacting particles and the consequent interacting states. If the surface roughness does not bring about substantial interaction due to mechanical interlocking, then increasing the surface roughness will reduce the interparticulate forces by means of reducing the effective contact area between the interacting particles (Iida *et al.*, 1993; Schaefer *et al.*, 1995). Increasing surface roughness of the interactive particles would also be expected to decrease particulate interaction due to liquid bridging since the adsorbed water layer will be more likely to be discontinuous on rougher surfaces than in the case of smoother surfaces (Coelho and Harnby, 1978). The measured surface tension has been reported to be much lower for particles with rough surfaces than particles with smoother surfaces under ambient conditions (Sutton, 1976). However, in pharmaceutical processing, any means of reducing interparticulate forces by roughening the particle surface would prove to be problematic since many pharmaceutical powders are in the range of 1 to 1000 μm in diameter. It is practically difficult, if not impossible, to engineer such fine particles to have optimum surface roughness. A high degree of surface porosity and roughness is often associated with increased frictional forces between the particles and this usually results in poor flow properties of the powder.

The effects of particle shape on performance have always been one of the most intriguing fields in pharmaceutical powder technology. Particle shape plays a role in determining nearly all the physical and, occasionally, chemical properties of particulate materials. For example, particle shape determines the powder friction coefficient (Harwood and Pilpel, 1969), angular griseofulvin having a larger coefficient of friction than rounded griseofulvin. Particle shape also plays an important part in powder compaction properties (Ridgway and Scotton, 1970) and strength of the resultant tablets. Tablets made from relatively more angular material usually have higher strength and this has been ascribed to increased particle interlocking between anisometric particles (Rupp, 1977). Particle shape also has an impact on the rheological properties of powders by affecting the interparticulate forces between the component particles. If particles take a shape that favours interparticulate interactions, then it is likely that such a shape is not favourable for powder flow or deaggregation but is favourable for compactibility.

Similar to the effects of surface smoothness, the influence of particle shape on interparticulate interactions is highly complicated. Whether a shape increases or decreases interparticulate forces is largely dependent upon how the component particles interact with one another. For example, if elongated particles are packed loosely, then their shape tends to reduce contact area and increase separation distance and consequently, reduce interparticulate forces. When such powders are subjected to compaction or undergo extensive mixing processes, the component particles will arrange themselves into the most stable state where they become closely packed to each other. Consequently, after processing in this manner, an elongated shape might be expected to increase particle contact area and reduce particle separation distance and, therefore, ultimately increase the interparticulate forces.

2.6 POLYMORPHISM AND CRYSTALLINITY

Pharmaceutical solids range from drug substances, many of which are organic crystals, to excipients, which can be either organic or inorganic crystals. Other excipients, such as microcrystalline cellulose, may be partially crystalline whilst polymeric excipients, such as PVP, may be predominantly amorphous. In the case of crystals, the constituent atoms or molecules arrange themselves repetitiously in a specific three-dimensional array, whereas in the case of amorphous particles, the atoms or molecules are randomly placed, as in a liquid. Amorphous materials are typically prepared either by rapid precipitation or cooling, or lyophilisation and usually have higher thermodynamic energy than the corresponding crystal forms. Upon storage, amorphous materials tend to convert to the more stable crystalline form. Such a transformation during bulk processing and/or storage, is one of the major disadvantages in the development of an amorphous form of a drug.

Many compounds crystallise in various crystal forms with different internal lattices and this phenomenon is called polymorphism. The various crystal forms of the same material occur as a result of the different crystallisation conditions. For example, lactose can be obtained in either two basic isomeric forms namely, α- and β-lactose (Figure 2.6), or as an amorphous form. α-Lactose monohydrate is obtained by crystallisation from a supersaturated solution at temperatures below 93.5°C, whereas β-lactose crystals are obtained at temperatures over 93.5°C (Nickerson, 1974), when no water is incorporated into the crystal lattice. Therefore, the crystals of β-lactose exist as a non-hygroscopic, anhydrous form in contrast to α-lactose, which can occur both as the monohydrate and as anhydrous α-lactose. Dehydration by thermal exposure or desiccation of α-lactose monohydrate can convert it into the anhydrous form. For example, the treatment of α-lactose monohydrate *in vacuo* at temperatures of 100–130°C can result in the production of a very hygroscopic product known as unstable anhydrous α-lactose (Itoh *et al.*, 1978). However, thermal treatment in a humid atmosphere at temperatures above 110°C, or desiccation with appropriate liquids, such as dry methanol, may result in a non-hygroscopic product, called stable anhydrous α-lactose (Itoh *et al.*, 1978). Amorphous lactose can be prepared by extensive mechanical grinding of the crystallised form (Morita *et al.*, 1984), by spray-drying (Fell and Newton, 1971) or by freeze-drying of solutions (Morita *et al.*, 1984).

A specific crystal form will exhibit different shapes according to the crystallisation conditions and the existence of varying morphological appearances is known as crystal habit. Different crystal habits occur due to different relative growth rates of 'faces' on the crystals despite the internal structure, i.e. the arrangement of atoms or molecules in the crystals, remaining the same. In general, if the growth is of equal rate in all directions, then a cubical-looking shape will be produced. If the growth is inhibited in one direction, then a plate will occur and if it is inhibited in two directions, then a needle will result. For example, α-lactose monohydrate has been observed to occur in a wide

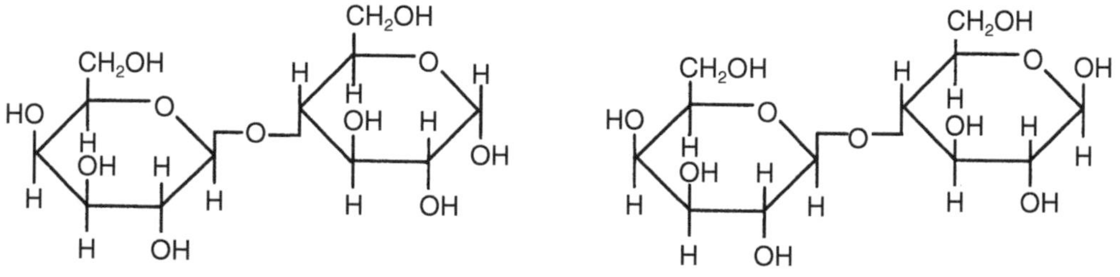

α-lactose

β-lactose

Figure 2.6: Chemical structures of different forms of lactose.

Figure 2.7: Diagrammatic representation of possible crystal habits of α-lactose monohydrate crystals.

variety of shapes, depending upon the conditions of crystallisation (Zeng *et al.*, 2000). The principal factor determining the crystal habit of lactose is supersaturation, i.e. the ratio of actual concentration to the solubility at a specific temperature (Herrington, 1934). At high supersaturation, crystallisation is forced to occur rapidly and only elongated prisms form (A or B in Figure 2.7). As the supersaturation decreases, the dominant crystal shape changes to diamond-shaped plates (C), then to pyramids (D), which result from an increase in the thickness of the diamond. Under conditions of slow crystallisation, lactose crystals usually exhibit the shape of tomahawks (E), this being the most common shape of lactose crystals.

These habit modifications arise due to both differences in the growth rates of the individual faces and changes in these relative growth rates with supersaturation (van Kreveld and Michaels, 1965; Michaels and van Kreveld, 1966). It has been shown that lactose crystals only grow in the direction of the principal axis and therefore such crystals have their nucleus in the apex of the tomahawk (van Kreveld and Michaels, 1965).

Significantly different polymorphs of a material besides being of different shape, may also have different physico-chemical properties such as melting point, density, hardness, optical properties and vapour pressure (Haleblian and McCrone, 1969). For example, various polymorphs and alternative 'faces' of a crystal form of lactose have been found to exhibit different degrees of hardness (Table 2.3) (Wong *et al.*, 1988).

In the case of α-lactose monohydrate crystals, the crystal face (011) has been shown to be harder, more elastic and have higher resilience than the crystal face (110). The face (011) of α-lactose crystals possesses the largest surface area, i.e. the face that grows the slowest during crystallisation and such differences in the mechanical properties of faces on a crystal may be attributed to the differences in growth rates. It has been widely accepted that molecules pack less densely on crystal faces that grow faster and the face that grows faster becomes smaller than the face that grows slower. Thus, the face (011) would be expected to have a higher molecular packing than the face (110). This may explain why face (011) was observed to be harder than face (110). Furthermore, the various faces of a crystal may possess different proportions of functional groups such that one face may be more hydrophilic than another. Therefore, if the crystal is subject to varying environmental humidities, then water vapour condensation may occur more favourably on more hydrophilic faces than on the more hydrophobic faces. Such

CHAPTER 2

TABLE 2.3

Some mechanical properties of monocrystals of α-lactose monohydrate and α-lactose anhydrate

Crystals	Crystal face (Miller index)	Hardness (MPa)	Elastic modulus (GPa)	Elastic quotient[a]
α-Lactose monohydrate	(011)	66.7	1.52	0.84
	(110)	43.3	0.83	0.80
β-Lactose anhydrous	(011)	29.9	0.87	0.44
	(110)	25.6	0.75	0.40

[a]The Elastic quotient index is the fraction of the indentation that behaves truly elastically and is a measure of the ability of materials to form tablets.
(data from Wong et al., 1988)

differences in affinity between different faces could also have significance for the adsorption to the surface of the other, smaller, particles such as drugs.

α-lactose monohydrate is harder and more elastic when compared with β-lactose. Anhydrous α-lactose has also been reported to have better compressibility than α-lactose monohydrate (Lerk et al., 1983). Due to their preferential properties in certain pharmaceutical processes such as tabletting, all the crystal forms of lactose mentioned above have been made commercially available (Kibbe, 2000).

The difference in the physico-chemical properties between the amorphous and crystalline portions of a material is often more pronounced than that between different polymorphs of the material. The constituent atoms or molecules are arranged randomly in an amorphous solid and, hence an amorphous material has higher internal and surface energy than the respective crystalline structure. Amorphous materials consequently have higher solubilities and lower melting points than crystalline solids. Furthermore, amorphous materials are softer and are more likely to undergo plastic deformation than crystalline particles. For example, spray-dried salbutamol sulphate was found to have a higher work of adhesion to median grade lactose crystals than micronised drug particles (Chawla et al., 1993). This may largely be due to the fact that micronised particles were harder than the spray-dried particles, since the latter are mostly composed of amorphous drug.

In reality, crystalline materials often possess some portions of disorder or amorphous regions within the crystals or on the surface. Crystallinity is often employed to assess the state of a material and it is often expressed as the percentage of the crystalline portion to the total quantity of a solid. The opposite term, amorphicity, is not unnaturally a term used to assess the amorphous content of a solid and it is expressed as the percentage of amorphous portion to the total solid. The amorphous portion within a crystal results from disarrangement of molecules in the crystal lattice during crystallisation. However, the amorphous region on a crystal surface may be produced by the stresses due to previous history (Buckton, 1995). Nearly every powder-handling process may impart some form of disorder to the crystal surface. Processes such as milling, spray-drying, compaction, lyophilisation, etc., of pharmaceutical solids, often induce at least partial conversion of most substances to a high energy form (Vadas *et al.*, 1991). Most noticeably, milling is capable of inducing amorphous regions on the crystal surfaces of many materials. For example, the crystallinity of cephalothin sodium has been shown to decrease to as low as 30% after milling for 2 h (Otsuka and Kaneniwa, 1990) and the longer a crystalline substance is subject to milling treatment, the less crystalline it will become (Nakagawa *et al.*, 1982). Introduced amorphous regions have been associated with enhanced chemical reactivity such as degradation (Nakagawa *et al.*, 1982) relative to the thermodynamically favoured crystalline state and they are referred to as 'hot' spots of the bulk solid. If these crystals are exposed to environmental water vapour, then water condensation may occur preferentially on these amorphous regions (Ahlneck and Zografi, 1990). Even relatively low percentages of amorphous material can adsorb considerable amounts of water into their structure and undergo considerable changes in physico-chemical properties, affecting the overall properties of the bulk substance. Therefore, any amorphous regions may 'magnify' the effects of water vapour condensation on the properties of the bulk materials. For example, if a material possesses 1% amorphous regions and it adsorbs 1% water, then it is likely that this moisture content is concentrated in such regions. Thus, this seemingly insignificant amount of moisture could crucially affect the physico-chemical properties in these regions and lead to a profound change in the overall properties of the bulk material. This is particularly the case for low molecular weight materials that may recrystallise readily due to their overall greater molecular mobility when compared to larger molecules such as polymers. Both higher surface energy or deformability of the amorphous regions may promote interparticulate forces between crystal surfaces. Preferential water condensation on these amorphous regions may also enhance particulate interaction through capillary forces. If recrystallisation occurs in such regions, then particulate interaction will be further promoted due to solid bridging.

2.7 PROCESSING CONDITIONS

Any external force that acts to increase particle deformation will undoubtedly increase the interparticulate forces. For example, the adhesion forces of micronised particles to a compacted disc were found to be determined by the press-on forces (Podczeck *et al.*, 1995a). When the applied forces increased from about 4.7×10^{-12} to 14.6×10^{-12} N, the adhesion forces of micronised lactose particles (3–5 μm) to a compacted lactose disc were found to increase from 5.09×10^{-12} to 11.12×10^{-12} N whereas the adhesion forces of micronised salmeterol xinafoate increased from 9.74×10^{-12} to 11.12×10^{-12} N. Although the adhesion of either micronised lactose to compacted salmeterol xinafoate or micronised salmeterol xinafoate to a compacted lactose disc increased as a function of applied pressure, lactose particles adhered more strongly to salmeterol xinafoate compacts than salmeterol xinafoate particles to lactose compacts. Theoretically, the adhesive forces should have remained the same whichever micronised particle was adhered to the opposing surface. This deviation from theory was attributed by the authors to a rougher surface of the compacted salmeterol xinafoate disc in comparison to the lactose disc (Podczeck *et al.*, 1995a). However, the use of compacted discs may not reflect the real situation existent within ordered mixes and the compacting process may also have changed the surface properties with regard to other parameters such as the free energy of the substrate surface. Thus, the adhesion of salmeterol xinafoate particles to lactose particles (200–250 µm) was compared with that of the same drug particle to a compacted lactose disc (Podczeck *et al.*, 1995b). For smaller press-on forces (i.e. $\leq 0.22 \times 10^{-8}$ N), adhesion of salmeterol xinafoate to lactose particles was shown to be similar to that of the drug to the lactose disc, but at higher press-on forces the adhesion force to compacted lactose surfaces was shown to be nearly twice as high as the adhesion to lactose particles.

The influence of duration of the applied pressure on particle adhesion was also investigated by Lam and Newton (1993). Particles of different hardness were initially pressed onto a steel surface by centrifugation and then adhesion forces were determined by re-centrifugation. All the particles investigated showed an increase in the adhesion forces to the substrate with time period of the adhesion (Figure 2.8).

Although the measurement of adhesion forces by centrifugation may reveal some characteristics of particulate interactions under the specific process conditions of the experiments, the results obtained by the centrifugation method can by no means be extrapolated to the more routinely encountered situation where most particulate interactions happen in a rotating container with neither surface compacted. The external forces acting on the interacting particles may arise from particle collision and shear forces due to particles intermingling. Consequently, it is practically impossible to calculate these forces and they may only be compared qualitatively. For example, increasing the rotating speed of a mixer may increase the forces and number of particle collisions, which may be analogous to an increase in the press-on forces between these particles. Thus, it is possible that even increasing the rotation speed of a mixer may increase the interparticulate forces within the mixed powder. Furthermore, during extensive blending, particles may seek to orientate in a position of smallest potential energy, which involves the shortest separation and largest contact area. This phenomenon might be of special relevance to the blending of elongated and plate-like particles. Thus, a higher adhesive force might be expected under these conditions. Kulvanich and Stewart (1988) showed that the adhesive tendency of drug particles in a model drug–carrier system increased with blending time. These effects may be partly due to the fact that under extensive blending, drug particles would progressively move to positions of greatest

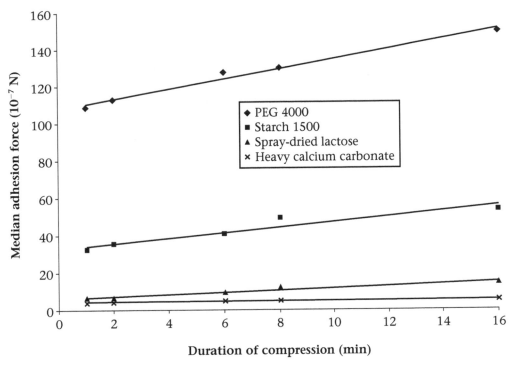

Figure 2.8: Relationship between the duration of compression and adhesion forces (adapted from Lam and Newton, 1993).

stability (Ganderton and Kassem, 1992), although increased triboelectrification of the particles may also partly contribute to the increased adhesion.

Hersey (1975) suggested that some areas of carrier surfaces are devoid of binding sites whilst other areas possess strong binding sites. On clean surfaces, these strong binding sites are available for adhesion and would induce strong forces between the carrier and adhered particles. However, on contaminated surfaces, these strong binding sites may be more likely to be saturated by the contaminants and thus more particles may be adhered to less strong binding sites and lower adhesive forces between the particles and carrier surface may be expected.

The existence of active binding sites on carrier particles has been supported by the findings of previous workers. Staniforth *et al.* (1981a) employed ultracentrifugation to measure the adhesion forces between salicyclic particles of mean radius of 2.5 µm and different carriers including recrystallised lactose. The adhesion profile was found to be largely influenced by the quantity of drug powder, increasing which generally decreased the median adhesion force. For example, the median adhesion force on recrystallised lactose decreased from 7.8×10^{-3} N to 2×10^{-3} N for 2% and 5% drug powder respectively and the percentage of fine particles with forces greater than 1.36×10^{-2} N was

18% in the former case but only 6% in the latter. These results suggest that increasing the amount of drug powder results in more drug particles being adhered to weak binding sites and hence the overall median adhesive force is reduced. Similar results were reported in the case of ordered mixes composed of potassium chloride powder (median diameter 5 μm) and carriers including Emdex (a spray-crystallised maltose-dextrose), Dipac (a direct compacting sugar) and recrystallised lactose (Staniforth and Rees, 1983). The stability of all the ordered mixes towards vibration decreased with an increase in the concentrations of potassium chloride, indicating that the overall adhesive forces were reduced.

2.8 FLOW PROPERTIES

One of the most important consequences of interparticulate forces in powder technology is that they can greatly affect powder flowability. These forces have a negative effect on powder flow in as much as the more they dominate the gravitational forces acting on the component particles, the worse the powder will flow. Therefore, all factors that affect interparticulate forces can influence powder flow.

2.8.1 Particle size

The importance of particle size in determining powder flow has long been recognised (e.g. Neumann, 1967). In general, the flow rate through an orifice increases rapidly with increasing size for fine particles until a maximum is reached, followed by a decrease as the particle size is increased further (Jones and Pilpel, 1966; Danish and Parrott, 1971). The decreased flow for small particles has been ascribed to the increasing importance of van der Waals, electrostatic and surface tension forces (Pilpel, 1964; Jones and Pilpel, 1966), while the decreased flow at larger particle sizes is attributable partly to limitations due to the size of the orifice relative to that of the particles. The tensile strength of a powder, which is inversely related to flow, was shown to be inversely proportional to the square of the particle diameter and directly proportional to the bonding force per point of contact (Rumpf, 1962). Therefore, the greater the interparticulate forces, the higher the tensile strength of the powder and the poorer the flow. As might be deduced, powder flow can be considered the opposite of powder agglomeration since flow is governed by gravitational forces whilst agglomeration is dependent upon interparticulate forces. As gravitational forces decrease with particle size more quickly than the interparticulate forces, the latter will become increasingly dominant over the former and hence particles will tend to agglomerate instead of flow. As discussed previously, for particles with a diameter <10 μm, van der Waals forces will dominate over gravitational forces. Therefore, such particles are often highly cohesive and have poor flowability.

2.8.2 Particle shape

Particle shape may affect the flow properties of a powder by changing interparticulate and/or frictional forces between the particles. In general, if the particle shape can reduce both interparticulate and frictional forces, then it is said to be favourable for good powder flow. A reduction in interparticulate forces by means of modifying particle shape alone cannot guarantee improved powder flow. For example, more angular materials have been found to have poorer flow than more rounded materials (Ridgway and Rupp, 1971). This was due to an increased friction coefficient for angular materials (griseofulvin) as compared with more rounded particles (Harwood and Pilpel, 1969), rather than to an increase in interparticulate forces for the more angular materials. Therefore, during tablet compression, more angular materials usually have a greater variation in die-fill weight than more rounded materials (Ridgway and Scotton, 1970). However, after being subjected to an external force, angular materials are more likely to interlock mechanically with each other when compared to rounded ones and, in fact, tablets made with the former were shown to have higher strength than those made with the latter (Rupp, 1977).

The effect of particle shape on powder flow is more complicated than particle size. Materials of different mechanical properties would need to possess a unique particle shape for optimised processing and this may change for different processing conditions. Therefore, it is difficult to generate a universal law describing the effects of particle shape on powder flow. However, sphericity is still the most sought after shape for the majority of pharmaceutical powders since this is believed to produce the least interparticulate friction and, hence, the best flow properties under normal conditions.

2.8.3 Electrostatic forces

Since most pharmaceutical solids do not conduct electricity, electrostatic charging occurs during nearly every powder-handling process such as milling, micronisation, flow, fluidisation, tabletting and capsule filling. Electrostatic charge can also arise from collisions between particles and the container walls. The triboelectrostatic charge is regarded as a nuisance in the majority of cases since it not only reduces powder flowability (Staniforth, 1982), especially in the case of fine particles, but also affects the bulk density of the powder by introducing electrostatic forces. In most of the cases, the electrostatic charge is non-reproducible and does not distribute homogeneously within a powder bed, so different regions may have varying densities and flowability and this could have a detrimental effect on the dose uniformity during, for example, tabletting and capsule filling. Triboelectrification is also one of the major reasons why powders stick to the walls of any contacting containers. Not surprisingly, electrostatic charge is often regarded as unwelcome. Although, in many cases, it is impossible to exclude completely any charge transfer during the handling of a powder, e.g. in milling, mixing, powder flow and drying operations, there are various ways to diminish or at least minimise the triboelectrification within the powder by means of improving the conductivity of the particle surface and the surrounding environment.

Moisture sorption on the particle surface is one of the most efficient ways to dissipate surface charge as the water increases the conductivity of both the particle surface and the surrounding environment. Therefore, the handling of a powder in a high relative humidity (e.g. >60%) is preferential in terms of diminishing the effects of triboelectrification on powder behaviour. However, as mentioned above, water sorption on the particle surface can increase interparticulate forces due to capillary forces. Too much water sorption on a particle surface is undesirable. Therefore, these two opposing factors have to be balanced carefully, such that sufficient charge dissipation is achieved without introducing excessive capillary forces within the powder. Since materials exhibit different properties of electrification and capillary condensation, the relative humidity under which they are processed should be optimised. Increasing the surface polarity of the component particles may also accelerate the charge dissipation accumulated on the particle surface. If there are polar groups in the chemical structure of a material, then, these groups should be present on the surface of a particle. Such a preferential orientation of polar groups may be achieved by controlling the crystallisation conditions of the material. There is a general rule that crystals of a material usually have higher surface polarity after crystallisation from a more polar solvent. Thus, crystallisation from such solvents is preferred if more polar crystals are to be produced providing that increasing the polarity of crystallisation solvent does not change other properties of the crystals such as introducing polymorphism or altering particle morphology. Particle surface polarity may also be modified by the addition of polar excipients that coat the material surface and thus, introduce polarity to the powder. Triboelectrostatic charge may also be diminished by treatment of the powder with ionised air, such as air that has passed through an electric field of high voltage.

Therefore, although there are many factors that influence the flowability of a powder, most do so through interparticulate forces. If they dominate over the gravitational forces acting on the component particles, then it is likely that the powder will exhibit poor flowability. Consequently, their reduction is likely to lead to improved powder flowability.

2.9 POWDER MIXING

Mixing may be defined as the intermingling of two or more dissimilar portions of material resulting in the attainment of a desired level of uniformity in the final product (Richardson, 1950). Mixing is among the oldest, most widely employed and, paradoxically, most neglected of pharmaceutical techniques. Since everyone from a person baking cakes to a chemist stirring the contents of a glass beaker with a rod has some familiarity with mixing procedures, it is often regarded as too obvious to require any serious academic study. However, powder mixing is a far more complicated process than has been appreciated since mixtures are composed of a number of particles that differ in density, size, shape, surface texture and mechanical properties. To obtain uniform mixing and sufficient stability of uniformity against any possible vibration during powder handling is the prerequisite for the preparation of many pharmaceutical solid dosage forms.

Powder mixing can be broadly classified either as randomised (Figure 2.9b) or ordered mixing (Figure 2.9c) (Hersey, 1975, 1979). In randomised mixing, the operation is carried out with particles of the same material, with identical size, shape and surface texture. Furthermore, there should be no interaction between the particles or between the particles and the container. In such a system, there will be an equal probability of a particle of either ingredient being found at any particular position in the mixture. Thus, the mixing homogeneity will obey the laws of probability and will be governed by the pure statistical distribution of the components. However, in reality, most pharmaceutical solids require the mixing of particles of different materials with different morphologies. When interparticulate forces of either attraction or repulsion are introduced within the mixtures, the randomised distribution of components will be disturbed and this type of mixing is best described by ordered mixing.

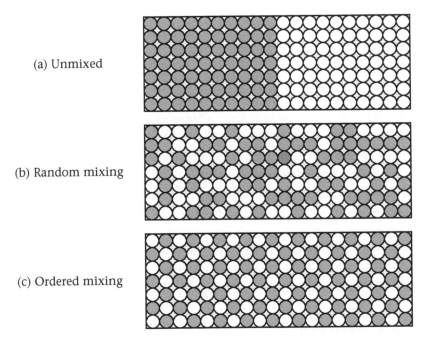

(a) Unmixed

(b) Random mixing

(c) Ordered mixing

Figure 2.9: Schematic diagram of powder mixing: (a) two ingredients before mixing; (b) randomised mixture of equal proportions of black and white particles; and (c) ordered mixture of equal proportions of black and white particles.

A typical example of an ordered mix is a two-component system where one component has a positive and the other a negative charge. Due to the interparticulate attraction of the two different components and the repulsion between particles of the same component, the material tends towards an alternating state of order in the case of a 1 : 1 mixture. Such mixing is apparently ordered and is more homogeneous than randomised mixing. Ordered mixing has also been proposed to be a 'particle crystalline state' (Fuhrer, 1996) since the state of order in an ordered mixture is similar to a polycrystalline state.

Pharmaceutical ordered mixes are mostly composed of fine drug particles and coarse carrier particles such as the mixtures between salicyclic acid and sucrose and those between magnesium stearate and tablet granules. Fine particles possess intrinsic cohesive properties but after these have been overcome such particles will have a tendency to adhere to the large particles of a second constituent such as a coarse carrier particle, due to the application of an external force as imposed by the mixing process. The uniformity of the ordered mixture is largely dependent upon both the cohesive forces acting between the fine particles, and the forces of adhesion of the fine particles to the coarse carrier particles. In order to obtain a uniform, stable ordered mixture, it is generally required that the adhesive forces of the fine particles to the coarse carrier particles exceed the cohesive forces of the fine particles. Otherwise, the fine particles will dislodge from the carrier particles during the vibrations encountered in nearly every powder-handling process, to form a heterogeneous mixture, a process called segregation (Staniforth and Rees, 1982a). Powder segregation also occurs as a result of differences in the physical characteristics of the materials, such as particle density and mechanical properties of the constituent particles. Although vibration is a major cause of segregation when powders are processed, increasing the interparticulate forces between the drug and carrier particles has proved to be one of the most efficient ways to minimise particle segregation within an ordered mixture. Several approaches have been employed to stabilise ordered mixes by means of increasing the interparticulate forces between the drug and carrier particles. These include selection of carrier particles with a large number of active binding sites in relation to accessible particle surface area (Staniforth *et al.*, 1981a), increasing the surface roughness of the carrier particles so as to increase interparticulate interaction between the drug and carrier particles due to mechanical interlocking (Staniforth *et al.*, 1981b), optimising the powder surface–electrical properties to introduce electrostatic forces between the carrier and drug particles (Staniforth and Rees, 1982b) and increasing capillary forces between the drug and carrier particles (Staniforth, 1985).

2.10   AGGLOMERATION

Agglomeration describes a size enlargement method by which fine particles are formed into larger entities by mechanical agitation, usually in the presence of a liquid phase.

Particle agglomeration engenders more favourable properties for further processing and hence, it is a widely employed technique in a variety of processes. For example, it may impart desirable flow properties, reducing dusting hazards and losses, facilitating the recovery of fines, preventing the segregation of components, securing controlled porosity, promoting fluidisation and creating a definite size and shape of particulate materials (Orr, 1966). In the pharmaceutical industry the most important objectives and benefits of agglomeration processes are to prepare powdered materials suitable for tabletting by creating non-segregating blends with improved flow and compaction properties (Kristensen, 1995). Fine particles can be agglomerated by wet granulation with a polymer binder solution or by melt granulation with a binder material that is softened or becomes molten at elevated temperatures but which returns to the solid state at room temperature.

The interparticulate forces between pairs of small particles must be strong enough for the agglomerates to withstand any external vibration during powder handling. These forces can be van der Waals, electrostatic, capillary or solid bridging. However, the strongest binding among particle masses is usually achieved when a distinct fusion occurs between points of contact, i.e. when solid bridging occurs. Solid bridging can be brought about by (1) partial or incipient melting which involves heating mixtures to a temperature below the melting point of the major constituent but above that of the lesser component, (2) solid diffusion, (3) added agents that set or harden, (4) chemical reactions between solids and/or the binding compounds and (5) crystallisation of dissolved materials (Rumpf, 1962).

Fusion among particles is usually brought about by melting the components at elevated temperatures and such a process is aided by high pressure and by impurities. Particulate fusion can be best obtained if eutectic compositions are employed. However, for some materials with high melting points, fusion can still be possible at the tips of surface roughness since the pressure on these contact areas may be sufficient to cause melting at these contact regions. Even if melting is not possible, fusion can still occur at the contact areas via diffusion, which is accelerated by high temperatures and pressures. Solid bridging is largely dependent upon particle deformability and hence plastic materials tend to form stronger solid bridges than elastic materials under similar conditions.

Wet granulation often involves the addition of a small amount of solvent, usually water, to the powdered material, resulting in an increase in interparticulate forces due to capillary action. These forces will then bind particles together to form larger agglomerates which are then dried to remove the binding liquid. The drying process may convert a relatively weak liquid bridge to a much stronger solid bridge via at least two mechanisms. First, wetting of the particle usually softens the particle surface, leading to a more plastic deformation at the contact areas of these particles. Second, some components may have dissolved in the liquid layer and subsequent drying will lead to the formation

CHAPTER 2

of solid bridges due to the precipitation and crystallisation of these materials. Finally, liquid bridging between particles can be achieved by either low or high viscosity liquids.

Whenever a particulate material is mixed with a liquid, the cohesive forces due to the capillary forces may cause agglomeration and consolidation of the agglomerates insofar as they can resist the disruptive forces by mechanical agitation of the moistened material. The mechanism by which the capillary forces cause particles to adhere to each other has been discussed in chapter 1. The interfacial tension peripheral to the liquid boundary draws the solids together, although the relative amount of liquid and its properties such as surface tension and viscosity, the particle morphology and wettability all determine the final interparticulate forces.

On the basis of the relative amount of the liquid incorporated within the mixture, liquid bridging in a powder can be classified into four states (Figure 2.10) (Newitt and Conway-Jones, 1958). The relative amount of liquid is expressed in terms of liquid saturation, i.e. the volume of liquid relative to the volume of pores and voids between solid particles. When the void spaces among the particles are only partially filled with the liquid, the pendular state is said to exist (Figure 2.10a). Here, discrete, lens-shaped liquid bridges are created among the particles. In this state, the liquid tension is directed along the surface and a negative capillary pressure is created within the bridge. Both effects result in a mutual pulling together of the particles. By increasing the liquid saturation, the funicular state is obtained (Figure 2.10b), where voids co-exist with a continuous network of liquid bridges. Further increasing the liquid saturation results in the capillary state (Figure 2.10c), where the void space of the powder is completely filled with liquid. In the capillary state, interfacial forces still exist and, insofar as the liquid does not extend to the particles' edges, there are concave cavities on the surface, a negative pressure exists internally, which further adds to the cohesive force between the particles. Further increasing the liquid saturation leads to the creation of a droplet state (Figure 2.10d), where the liquid completely envelops the particles. In this state, the concavities due to pores are replaced by the convex surface of a continuous liquid drop. All interparticulate forces vanish and only the interfacial tension of the liquid tends to hold the particles together. A liquid saturation of about 25–35% usually results in the pendular

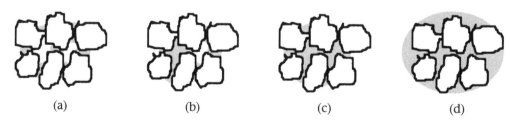

(a)   (b)   (c)   (d)

Figure 2.10: States of liquid bridging in a powder (a) pendular state, (b) funicular state, (c) capillary state and (d) droplet state.

and funicular states whilst saturation of about 80% initiates the capillary state (Capes, 1980).

The tensile strength, σ_t, of an agglomerate is a function of the cohesive forces F_c at the contact points and an inverse function of both the particle diameter d, and the porosity of the agglomerate, ϵ. For a simplifying agglomerate which is composed of monosized spherical particles, the tensile strength can be calculated as (Rumpf, 1962):

$$\sigma_t = \frac{9}{8} \frac{(1 - \epsilon)}{\epsilon} \frac{F_c}{d^2} \qquad (2.5)$$

where σ_t is the mean tensile strength per unit section area and F_c is the cohesive force of a single bond.

If the liquid saturation is such that a pendular state is obtained in the powder, the particles are in close contact and the contact angle of the liquid is zero, the cohesive force of a single bond can be approximated as:

$$F_c = 2\pi\gamma d \qquad (2.6)$$

where γ is the surface tension of the liquid.

Equation (2.5) can then be simplified as (Pietsch, 1969):

$$\sigma_t = \frac{9}{4} \frac{(1 - \epsilon)}{\epsilon} \frac{\gamma\pi}{d} \qquad (2.7)$$

Thus, the tensile strength of an agglomerate in the pendular state is more or less a constant, regardless of the liquid saturation.

The potential of water to pull particles together upon evaporating from the voids of an agglomerate can be measured using a bed of the powder (Elzinga and Banchero, 1961). If a powder bed in a capillary state is formed in a vertical container such as a glass tube, which is connected distally to a manometer, a considerable reduction in the pressure will be measured as the liquid evaporates from the upper levels of the bed. The suction will increase as moisture is progressively vaporised until a value designated as the entry suction is attained. This point is reached when air begins to enter the voids or pores between the particles, i.e. the funicular state. Thereafter, the suction will remain more or less constant until the pendular state is approached.

The entry suction or capillary pressure is a measure of the forces tending to pull particles together. The higher the capillary pressure, the stronger the binding forces between particles in a capillary state which can be expressed (Rumpf, 1962) by:

$$P_s = \frac{(1 - \epsilon)}{\epsilon} S\gamma \cos\theta \qquad (2.8)$$

where P_s is the capillary pressure, S is the specific surface area, γ is the surface tension of the liquid and θ is the contact angle of the liquid on the surface.

For uniform spherical particles of diameter d, S is equal to $6/d$ and then equation (2.8) can be converted to:

$$P_s = 6 \frac{(1 - \epsilon)}{\epsilon} \frac{\gamma}{d} \cos \theta \qquad (2.9)$$

Apart from the capillary pressure, the interfacial tension of the liquid along the edge of the agglomerate also contributes to the overall tensile strength, which can thus be accounted for as:

$$\sigma_t = P_s + \frac{C}{A} \gamma \cos \theta = \left[\frac{6}{d} \frac{(1 - \epsilon)}{\epsilon} + \frac{C}{A} \right] \gamma \cos \theta \qquad (2.10)$$

where C is the agglomerate circumference and A its cross-sectional area.

According to equations (2.7) and (2.9), a capillary-state agglomerate would be expected to be some two to three times stronger than a pendular one of similar components.

Because the funicular state is intermediate between the pendular and capillary states, a funicular-state agglomerate should exhibit a tensile strength ranging between those of the pendular and capillary state, which can be approximated as:

$$\sigma_t = S_{liq} K \frac{(1 - \epsilon)}{\epsilon} \frac{\gamma}{d} \cos \theta \qquad (2.11)$$

where S_{liq} is the liquid saturation and K is a constant determined by the shape of the component particles in the agglomerate. K has a value of 6 for uniform spherical component particles but is between 6.5 and 8 for irregular sand particles (Capes, 1980).

The above equations were generated assuming monosized, spherical particles. However, these equations may also be employed to estimate the tensile strength of agglomerates composed of poly-dispersed particles, where the volume–surface diameter d_{vs} of the component particles is used to replace the particle diameter d in the above equations (Kristensen, 1995). From these equations, it can be seen that the tensile strength of the agglomerate is largely dependent upon its porosity: the higher the porosity then the lower the tensile strength. Tensile strength is also a function of both the surface tension γ and contact angle θ. The higher surface tension of the liquid and/or smaller contact angle of the liquid onto the particle surface produces higher tensile strength of the agglomerates.

It is often believed that increasing the viscosity of the liquid will result in an increase in the tensile strength of the agglomerates (Bowden and Tabor, 1964). This is because the adhesion between the binding medium and the solid is almost always greater than the cohesive strength of the binder, possibly due to particle–surface irregularities and the physical interlocking between the medium and the solid particles. Therefore, liquids employed as binders generally exhibit high viscosity. These binders are dissolved in a solvent to facilitate dispersion throughout the powder and after solvent evaporation, the binders will then become intertwined with the component particles and hold them together.

Although van der Waals and/or electrostatic forces may play a role in determining the tensile strength of agglomerates, these forces are usually several orders of magnitude weaker than those produced by any of the previously described binding mechanisms.

REFERENCES

AL-ANGARI, A.A., KENNERLEY, J.W. and NEWTON, J.M. (1985) The compaction properties of polyethylene glycols. *J. Pharm. Pharmacol.* **37**, 151–153.

AHLNECK, C. and ZOGRAFI, G. (1990) The molecular basis of moisture effects on the physical and chemical stability of drugs in the solid state. *Int. J. Pharm.* **62**, 87–95.

BAILEY, A.G. (1984) Electrostatic phenomena during powder handling. *Powder Technol.* **37**, 71–85.

BASSAM, F., YORK, P., ROWE, R.C. and ROBERTS, R.J. (1990) Young's modulus of powders used as pharmaceutical excipients. *Int. J. Pharm.* **64**, 55–60.

BOWDEN, F.P. and TABOR, D. (1964) *The Friction and Lubrication of Solids*, Part 1. Oxford, Oxford University Press, 299–306.

BUCKTON, G. (1995) *Interfacial Phenomena in Drug Delivery and Targeting.* Chichester, Harwood Academic Publishers, 27–58.

BYRN, S.R. (1982) *Solid State Chemistry of Drugs.* New York, Academic Press.

CAPES, C.E. (1980) *Particle Size Enlargement.* Amsterdam, Elsevier.

CARSTENSEN, J.T. (1993) *Pharmaceutical Principles of Solid Dosage Forms.* Pennsylvania, Technomic Publishing Company, Inc.

CARTER, P.A., ROWLEY, G., FLETCHER, E.J. and HILL, E.A. (1992) An experimental investigation of triboelectrification in cohesive and non-cohesive pharmaceutical powders. *Drug Dev. Ind. Pharm.* **18**, 1505–1526.

CHAN, S.Y. and PILPEL, N. (1983) Absorption of moisture by sodium cromoglycate and mixtures of sodium cromoglycate and lactose. *J. Pharm. Pharmacol.* **35**, 477–481.

CHAWLA, A., BUCKTON, G., TAYLOR, K.M.G., NEWTON, J.M. and JOHNSON, M.C.R. (1993) Surface energy determinations of powders for use in dry powder aerosol formulations. *J. Pharm. Pharmacol.* **45**(Suppl.) p. 1148.

COELHO M.C. and HARNBY, N. (1978) The effect of humidity on the form of water retention in a powder. *Powder Technol.* **20**, 197–200.

■ CHAPTER 2 ■

DANISH, F.Q. and PARROTT, E.L. (1971) Flow rates of solid particulate pharmaceuticals. *J. Pharm. Sci.* **60**, 548–554.

ELZINGA JR., E.R. and BANCHERO, J.T. (1961) The mechanics of drops in liquid–liquid systems. *Am. Inst. Chem. Eng. J.* **7**, 394–399.

FELL, J.T. and NEWTON, J.M. (1971) The production and properties of spray dried lactose, part 3: the compaction properties of samples of spray dried lactose produced on an experimental drier. *Pharm. Act. Helv.* **46**, 441–447.

FUHRER, C. (1996) Interparticulate attraction mechanisms. In: ALDERBORN, G. and NYSTROM, C. (eds), *Pharmaceutical Powder Compaction Technology*. New York, Marcel Dekker, 1–15.

FUKUOKA, E., KIMURA, S., YAMAZAKI, M. and TANAKA, T. (1983) Cohesion of particulate solids. VI. Improvement of apparatus and application to measurement of cohesiveness at various levels of humidity. *Chem. Pharm. Bull.* **31**, 221–229.

GANDERTON, D. and KASSEM, N.M. (1992) Dry powder inhalers. In: GANDERTON, D. and JONES, T. (eds), *Advances in Pharmaceutical Sciences*, Vol. 6. London, Academic Press, 165–191.

HALEBLIAN, J. and MCCRONE, W. (1969) Pharmaceutical applications of polymorphism. *J. Pharm. Sci.* **58**, 911–929.

HARWOOD, C.F. and PILPEL, N. (1969) The flow of granular solids through circular orifices. *J. Pharm. Pharmacol.* **21**, 721–730.

HERRINGTON, B.L. (1934) Some physico-chemical properties of lactose. II. Factors influencing the crystalline habit of lactose. *J. Dairy Sci.* **17**, 533–542.

HERSEY, J.A. (1975) Ordered mixing: A new concept in powder mixing practice. *Powder Technol.* **11**, 41–44.

HERSEY, J.A. (1979) The development and applicability of powder mixing theory. *Int. J. Pharm, Tech & Prod. Mfr.* **1**, 6–13.

ICHIKAWA, J., IMAGAWA, K. and KANENIWA, N. (1988) The effect of crystal hardness on compaction propensity. *Chem. Pharm. Bull.* **36**, 2699–2702.

IIDA, K., OTSUKA, A., DANJO, K. and SUNADA, H. (1993) Measurement of the adhesive force between particles and a substrate by means of the impact separation method — Effect of the surface-roughness and type of material of the substrate. *Chem. Pharm. Bull.* **41**, 1621–1625.

ITOH, T., KATOH, M. and ADACHI, S. (1978) An improved method for the preparation of crystalline β-lactose and observations on the melting point. *J. Dairy Res.* **45**, 363–371.

JETZER, W., LEUENBERGER, H. and SUCKER, H. (1983) The compressibility and compactibility of pharmaceutical powders. *Pharm. Tech.* **7**, 33–39.

JONES, T.M. and PILPEL, N. (1966) The flow properties of granular magnesia. *J. Pharm. Pharmcol.* **18**, 81–93.

KIBBE, A.H. (2000) *Handbook of Pharmaceutical Excipients*, 3rd edn. Washington, American Pharmaceutical Association and Pharmaceutical Press.

KONTNY, M.J. and ZOGRAFI, G. (1985) Moisture sorption kinetics for water-soluble substances.4. studies with mixtures of solids. *J. Pharm. Sci.* **74**, 124–127.

KRISTENSEN, H.G. (1995) Particle agglomeration. In: GANDERTON, D., JONES, T. and MCGINITY, J. (eds), *Advances in Pharmaceutical Sciences*, Vol. 7. London, Academic Press, 221–272.

KULVANICH, P. and STEWART, P.J. (1988) Influence of relative humidity on the adhesive properties of a model interactive system. *J. Pharm. Pharmacol.* **40**, 453–458.

LACHMAN, L., LIEBERMAN, H.A. and KANIG, J.L. (1986) *The Theory and Practice of Industrial Pharmacy*, 3rd edn. Philadelphia, Lea and Febiger.

LAM, K.K. and NEWTON, J.M. (1991) Investigation of applied compression on the adhesion of powders to a substrate surface. *Powder Technol.* **65**, 167–175.

LAM, K.K. and NEWTON, J.M. (1993) The influence of the time of application of contact pressure on particle adhesion to a substrate surface. *Powder Technol.* **76**, 149–154.

LERK, C.F., ANDREAE, A.C., DE BOER, A.H., BOLHUIS, G.K., ZUURMAN, K., DE HOOG, P., KUSSENDRAGER, K. and VON LEVERINK, J. (1983) Increased binding capacity and flowability of α-lactose monohydrate after dehydration. *J. Pharm. Pharmacol.* **35**, 747–748.

LEUENBERGER, H. (1982) The compressibility and compactibility of powder systems. *Int. J. Pharm.* **12**, 41–55.

LEUENBERGER, H. and ROHERA, B.D. (1985) Compaction equations for binary powder mixtures. *Pharm. Act. H.* **60**, 279–286.

MARSH, D.M. (1964) Plastic flow in glass. *P. Roy. Soc. A* **279**, 420–435.

MICHAELS, A.S. and VAN KREVELD, A. (1966) Influences of additives on growth rates in lactose crystals. *Neth. Milk D.* **20**, 163–181.

MORGAN, B.B. (1961) Adhesion and cohesion of fine particles. *British Coal Utilisation Research Association, Monthly Bulletin*, **25**, 125–164.

MORITA, M., NAKAI, Y., FUKUOKA, E. and NAKAJIMA, S. (1984) Physicochemical properties of crystalline lactose. II. Effect of crystallinity on mechanical and structural properties. *Chem. Pharm. Bull.* **32**, 4076–4083.

NAKAGAWA, H., TAKAHASHI, Y. and SUGIMOTO, I. (1982) The effects of grinding and drying on the solid state stability of sodium prasterone sulfate. *Chem. Pharm. Bull.* **30**, 242–248.

NEUMANN, B.S. (1967) The flow properties of powders. In: BEAN, H.S., BECKETT, A.H. and CARLESS, J.E. (eds). *Advances in Pharmaceutical Sciences*, vol. 2. London, Academic Press, 181–221.

NEWITT, D.M. and CONWAY-JONES, J.M. (1958) A contribution to the theory and practice of granulation. *Trans. Inst. Chem. Eng. (London)* **36**, 422–440.

NICKERSON, T.A. (1974) *Fundamentals of Dairy Chemistry*, 2nd edn. Westport, CT, AVI Publishing Co. Inc., 300.

ORR JR., C. (1966) Agglomeration. In: *Particulate Technology*. New York, The Macmillan Company, 400–455.

OTSUKA, M. and KANENIWA, N. (1990) Effect of grinding on the crystallinity and chemical stability in the solid state of cephalothin sodium. *Int. J. Pharm.* **62**, 65–73.

PIETSCH, W.B. (1969) Strength of agglomerates bound by salt bridges. *Can. J. Chem. En.* **47**, 403–409.

PILPEL, N. (1964) The flow properties of magnesia. *J. Pharm. Pharmcol.* **16**, 705–716.

PODCZECK, F., NEWTON, J.M. and JAMES, M.B. (1995a) Adhesion and autoadhesion measurements of micronized particles of pharmaceutical powders to compacted powder surfaces. *Chem. Pharm. Bull.* **43**, 1953–1957.

PODCZECK, F., NEWTON, J.M. and JAMES, M.B. (1995b) Assessment of adhesion and autoadhesion forces between particles and surfaces. Part III. The investigation of adhesion phenomena of salmeterol xinafoate and lactose monohydrate particles in particle-on-particle and particle-on-surface contact. *J. Adhes. Sci. Technol.* **9**, 475–486.

RICHARDSON, E.G. (1950) The formation and flow of emulsions. *Journal of Colloid Science* **5**, 404–413.

RIDGWAY, K. and RUPP, R. (1971) The mixing of powder layers on a chute: the effect of particle size and shape. *Powder Technol.* **4**, 195–202.

RIDGWAY, K. and SCOTTON, J.B. (1970) The effect of particle shape on the variation of fill of a tabletting die. *J. Pharm. Pharmacol.* **22**, 24S.

ROBERTS, R.J. and ROWE, R.C. (1987) The Young's modulus of pharmaceutical materials. *Int. J. Pharm.* **37**, 15–18.

ROBERTS, R.J., ROWE, R.C. and YORK, P. (1991) The relationship between Young's modulus of elasticity of organic-solids and their molecular-structure. *Powder Technol.* **65**, 139–146.

ROWE, R.C. and ROBERTS, R.J. (1995) The mechanical properties of powders. In: GANDERTON, D., JONES, T. and McGINITY, J. (eds), *Advances in Pharmaceutical Sciences*, Vol. 7. London, Academic Press, 1–62.

RUMPF, H. (1962) The strength of granules and agglomerates. In: KNEPPER, W.A. (ed.), *Agglomeration.* New York, Wiley Interscience, 379–418.

RUPP, R. (1977) Flow and other properties of granulates. *Boll. Chim. Farm.* **116**, 251–266.

SCHAEFER, D.M., CARPENTER, M., GADY, B., REIFENBERGER, R., DEMEJO, L.P. and RIMAI, D.S. (1995) Surface-roughness and its influence on particle adhesion using atomic-force techniques. *J. Adhes. Sci. Technol.* **9**, 1049–1062.

SHOTTON, E. and HARB, N. (1966) The effects of humidity and temperature on cohesion of powders. *J. Pharm. Pharmacol.*, **18**, 175–178.

STANIFORTH, J.N. (1982) The effect of frictional charges on flow properties of direct compression tableting excipients. *Int. J. Pharm.* **11**, 109–117.

STANIFORTH, J.N. (1985) Ordered mixing or spontaneous granulation? *Powder Technol.* **45**, 73–77.

STANIFORTH, J.N. and REES, J.E. (1982a) Effect of vibration time, frequency and acceleration on drug content uniformity. *J. Pharm. Pharmacol.* **34**, 700–706.

STANIFORTH, J.N. and REES, J.E. (1982b) Electrostatic charge interactions in ordered powder mixes. *J. Pharm. Pharmacol.* **34**, 69–76.

STANIFORTH, J.N. and REES, J.E. (1983) Segregation of vibrated powder mixes containing different concentrations of fine potassium chloride and tablet excipients. *J. Pharm. Pharmacol.* **35**, 549–554.

STANIFORTH, J.N., REES, J.E., LAI, F.K. and HERSEY, J.A. (1981a) Determination of interparticulate forces in ordered powder mixes. *J. Pharm. Pharmacol.* **33**, 485–490.

STANIFORTH, J.N., REES, J.E., KAYES, J.B., PRIEST, R.C. and COTTERILL, N.J. (1981b) The design of a direct compression tableting excipient. *Drug Dev. Ind. Pharm.* **7**, 179–190.

SUTTON, H.M. (1976) Flow properties of powders and the role of surface character. In: PARFITT, G.D. and SING, K.S.W. (eds), *Characterization of Powder Surfaces with Special Reference to Pigments and Fillers.* London, Academic Press, 122.

VADAS, E.B., TOMA, P. and ZOGRAFI, G. (1991) Solid-state phase-transitions initiated by

CHAPTER 2

water-vapor sorption of crystalline L-660,711, a leukotriene d4 receptor antagonist. *Pharmaceut. Res.* **8**, 148–155.

VAN KREVELD, A. and MICHAELS, A.S. (1965) Measurement of crystal growth of alpha-lactose. *J. Dairy Sci.* **48**, 259–265.

WONG, D.Y.T., WRIGHT, P. and AULTON, M.E. (1988) The deformation of alpha-lactose monohydrate and anhydrous alpha-lactose monocrystals. *Drug Dev. Ind. Pharm.* **14**, 2109–2126.

YORK, P. (1978) Particle slippage and rearrangement during compression of pharmaceutical powders. *J. Pharm. Pharmacol.* **30**, 6–10.

ZENG, X.M., MARTIN, G.P., MARRIOTT, C. and PRITCHARD, J. (2000) The influence of crystallization conditions on the morphology of lactose intended for use as a carrier for dry powder aerosols. *J. Pharm. Pharmacol.* **52**, 633–643.

ZOGRAFI, G. and KONTNY, M.J. (1986) The interactions of water with cellulose-derived and starch-derived pharmaceutical excipients. *Pharmaceut. Res.* **3**, 187–194.

CHAPTER

Medicinal Aerosols

Contents

3.1   INTRODUCTION

The term 'aerosol' refers to an assembly of liquid or solid particles suspended in a gaseous medium long enough to enable observation or measurement. Aerosols may also be described in other non-scientific terms, according to the source or appearance of the particles. For example, liquid particles formed by condensation of supersaturated vapours or by physical shearing of liquids, such as nebulisation, spraying, or bubbling can commonly be termed 'fogs' or 'mists', whilst an aerosol consisting of solid and liquid particles, created partly by the action of sunlight on atmospheric water vapour is known as a 'smog'. Another example is the solid or liquid aerosol resulting from incomplete combustion, which is termed 'smoke'. Since aerosol particles range in diameter from about 10^{-9} to about 10^{-4} m, the unit of micrometer (1 μm = 10^{-6} m) is generally used when referring to the particle size of aerosols. For instance, a particle most hazardous to the human respiratory tract is of the order of 10^{-6} m in diameter and this is conveniently described as a 1-μm particle. The term 'micron' is sometimes unofficially used to refer to a micrometer.

Aerosols can be broadly classified into desirable and undesirable aerosols. Radioactive aerosols and industrial aerosols in the atmosphere are considered undesirable aerosols since, after inhalation into the respiratory tract, such aerosols are known to cause numerous health problems. More recently, various natural and man-made aerosols, as a by-product of industrialisation, have been shown to be one of the major reasons for global warming. Atmospheric aerosol particles also cause many problems in the production of high-speed integrated circuits, and hence the manufacture of such items needs to take place in an ultra-clean environment.

Among those considered to be desirable aerosols are medicinal aerosols, which may be defined as man-made aerosols that serve to improve the health of human beings. Medicinal aerosols can be traced back over 4000 years (Grossman, 1994). Early medicinal aerosols were mainly volatile aromatics, such as menthol, thymol and eucalyptus and 'smokes' generated from burning the leaves of *Atropa belladonna* and *Datura stramonium*. In the late 1900s, a glass bulb nebuliser was developed to generate a mist from stramonium (Grossman, 1994). Such a device was improved in 1938, when a hand-held squeeze-bulb nebuliser was developed to deliver adrenaline, atropine and papaverine (Butterworth, 1986). However, the use of this device was so laborious that frequent application caused asthmatics to develop hard calluses on the hand due to constant pumping (Bell, 1992). Since then, the design of nebuliser devices has undergone drastic refinement. Modern nebulisers can generate pre-designed medicinal aerosols via mechanical or sonic energies. However, the most striking development in medicinal aerosols was the invention by George Maison and Irving Porush of the portable gas-pressurised metered-dose inhaler (MDI) in 1955 (Theil, 1996). MDIs have dramatically improved the portability and usability of medicinal aerosols and thus greatly increased the popularity of medicinal aerosols, although these devices are far from perfect. Among their drawbacks are the need for co-ordination between actuation and inhalation but the most detrimental effect associated with the use of MDIs arises from the chlorofluoro-carbon (CFC) propellants. CFCs are notorious for their ability to deplete the ozone layer in the stratosphere, which protects the population from exposure to excessive UV radiation, known to cause skin cancer, disruption of ecological processes and global warming. The international community has agreed to phase out CFCs as soon as practicable. Therefore, alternative propellants or devices have to be developed to replace CFCs in MDIs. This has prompted the rapid development of another medicinal aerosol, the dry powder inhaler (DPI). DPIs offer a means of delivering the medication without using any propellants, since the dispersion and entrainment of drug particles are caused by the inhalation efforts of the patient alone. Thus, DPIs are environmentally friendly and very easy to use and handle, overcoming the problems associated with the synchronisation of actuation and inhalation during the operation of MDIs. DPIs are highly portable and relatively inexpensive in comparison with nebulisers. The drugs are kept in the solid state in DPIs and thus would be expected to have a high physico-chemical stability.

Up to now, medicinal aerosols have been employed to deliver drugs for mainly localised effects although there is much current interest in employing the lungs as a portal of entry into the body for inducing systemic effects. This route of drug administration provides a means of delivering drugs to their sites of action and hence maximised pharmacological effects can be obtained with minimised associated side-effects. This is particularly true for the many drugs used in the treatment of lung diseases such as bronchodilators, antiasthmatic drugs, antibiotic and antiviral agents. The direct delivery of drugs such as these to their target sites also produces a rapid and predictable onset of drug action. Furthermore, the human lung possesses a relatively large surface area in the order of 126 m^2 for drug absorption (Wall, 1995), lower metabolic activity when compared with the liver or gastrointestinal tract, desirable permeability of alveolar epithelium to many molecules as well as a rich blood supply. The lung may thus provide an ideal portal for drug delivery to the systemic circulation and as such has attracted special interest for the possible delivery of peptides, polypeptides and other macromolecules (Byron and Patton, 1994).

CHAPTER 3

3.2 AEROSOL DEPOSITION IN THE RESPIRATORY TRACT

Deposition is a process by which inhaled particles separate from the flow streamlines and contact a respiratory surface from which there is no rebound or resuspension. Particles that remain airborne throughout the respiratory cycle are exhaled. The deposition of particles in the airways is a very complicated process. Inhaled particles may be deposited at or captured in any site of the respiratory tract by different mechanisms,

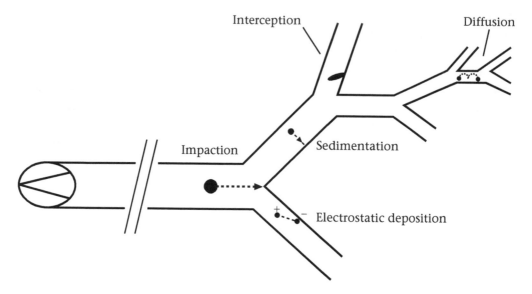

Figure 3.1: Schematic diagram showing deposition mechanisms of medicinal aerosols in the respiratory tract.

depending on the physiological factors, inhalation technique and physico-chemical properties of the particles. There are five possible mechanisms by which significant deposition may occur, namely, impaction, sedimentation, diffusion, interception and electrostatic precipitation (Figure 3.1).

3.2.1   Inertial impaction

Impaction is the most important mechanism of particle deposition. The inhaled air follows a tortuous path through the branching airways and each time the air stream changes its direction, the airborne particles tend to maintain their established trajectories. If the particles possess sufficient momentum, they will not follow the gas streamlines and will impact on any obstacles in their path.

If a particle of mass, m, is moving with velocity, V, through still air, the particle will stop after travelling a distance, S, as a result of frictional forces, in accordance with equation (3.1):

$$S = BmV \tag{3.1}$$

where B is the mobility of the particle, i.e. the velocity per unit of force. Thus, the greater the particle mobility, mass and velocity, the longer it will continue to fly in its original trajectory and, hence, it is more likely for the particle to hit any obstacle in the airways. That is to say the chances of deposition by impaction are increased.

For a particle travelling in an airway, the probability of its deposition by impaction is a function of a dimensionless parameter involving the air velocity, known as Stokes' number (Stk), which can be obtained, using equation (3.2):

$$Stk = \frac{\rho_p d^2 V}{18 \eta R} \tag{3.2}$$

where ρ_p is the particle density, d is the particle diameter, V is the air velocity, η is the air viscosity and R is the airway radius.

The higher the value of Stokes' number, the more readily particles will deposit by impaction. Therefore, increasing either the particle size or air flow rate can increase deposition by impaction. This is usually undesirable for most inhaled drug particles as impaction occurs most frequently in the upper airways which may not be the sites of action for most drugs. Therefore, with large particles, whenever the convective airflow is fast or turbulent as in the upper airways or changing direction at bifurcations between successive airways generations, impaction is the most likely mechanism of particle deposition. Hence, 'hot spots' are found particularly at bifurcations and on the carinal ridges (Martonen and Yang, 1996). Inertial impaction causes the major portion of aerosol deposition on a mass basis, and occurs mainly in the upper bronchial region (Hind, 1982).

3.2.2 Sedimentation

Sedimentation is a time-dependent process, in which particles travelling through an airway settle under the influence of gravity. An airborne particle with a density (ρ_p) greater than that of air (ρ_a), will experience a downward force (F) which can be expressed as:

$$F = \frac{\pi}{6} d^3 g(\rho - \rho_a) \tag{3.3}$$

The particle accelerates downwards and reaches a constant terminal velocity when the resistant force due to interaction with air molecules balances its weight. For spherical particles, Stokes' law can be used to predict the resistant force F_r:

$$F_r = 3\pi d \eta V \tag{3.4}$$

When $F = F_r$, the terminal velocity, V_t, is given by Stokes' law in the following form:

$$V_t = \frac{(\rho_p - \rho_a) d^2 g}{18 \eta} \tag{3.5}$$

Stokes' law is valid for unit density particles of 1 to 40 μm in diameter settling in air. For particles of larger diameter, the resistant force is more than that predicted by Stokes' law as a result of the inertial effects of accelerating a mass of gas to push it aside. When particle size decreases, the resistance force is less than that predicted by Stokes' law since the particle size is comparable in size to the mean free path of air, λ, (normally 0.0665 µm) molecules and can slip between other molecules. A correction factor, known as the Cunningham slip correction, must be applied to the calculation to obtain the true value (Allen and Raabe, 1985). Further, equation (3.5) assumes laminar flow within the airways. However, as the air flow increases, the stream will become turbulent, characterised by large and irregular velocity fluctuations. The flow pattern is expressed by the dimensionless Reynolds number, R_e (equation 3.6):

$$R_e = \frac{\rho_g V d'}{\eta} \tag{3.6}$$

where V is the velocity of the gas, η the dynamic gas viscosity, ρ_g the gas density and d' a characteristic dimension of the object. Laminar flow occurs in a circular duct when the flow R_e is less than about 2000, whilst turbulent flow occurs for R_e above 4000 (Baron and Willeke, 1992). Turbulent flow will increase the contact of particles with the airways and hence, increase the deposition by impaction.

Sedimentation is dominant in particle deposition in the smaller airways and alveolar region, where flow velocities are low and airway dimensions are small (Hind, 1982). Aerosol particles tend to settle onto airway surfaces most efficiently either during slow, steady breathing or during breath-holding (Hatch and Gross, 1964). These breathing manoeuvres allow a sufficiently long residence time of particles within the lungs for sedimentation to occur.

When particle size diminishes, the motion of the particle becomes primarily diffusive, due to the molecular bombardment, known as Brownian motion.

3.2.3 Diffusion

Particles less than 1 µm in diameter will undergo a random motion, Brownian motion, in the air stream due to the bombardment of the gas molecules. Brownian motion increases with decreasing particle size and becomes an important mechanism for particle deposition in the deep lung.

The root mean square displacement, Δ, with time, t, for Brownian motion, is expressed as:

$$\Delta = \sqrt{6Dt} \tag{3.7}$$

where D is the diffusion coefficient of the particle and

$$D = KT/3\pi\eta d \tag{3.8}$$

TABLE 3.1

Particle displacement (μm) in 1 s under conditions similar to the respiratory tract (adapted from Brain *et al.*, 1985)

Particle diameter (μm)	Displacement (μm) by	
	Brownian motion	Sedimentation
50	1.7	70,000
20	2.7	11,500
10	3.8	2900
5	5.5	740
2	8.8	125
1	13	33
0.5	20	9.5
0.2	37	2.1
0.1	60	0.81
0.05	120	0.35

where K is the Boltzmann constant, T is the absolute temperature, η is the gas viscosity and d is the particle diameter.

Thus, particle size is the key factor in determining the mechanism of particle deposition whether this is due to sedimentation or diffusion. The displacement in 1 s of a unit density sphere due to gravitational sedimentation and Brownian motion for particles from 0.05 to 50 μm is shown in Table 3.1.

Thus, Brownian motion increases with decreasing particle size whilst sedimentation exhibits the inverse trend. The total displacement decreases with decreasing particle size until it reaches a minimum value at about 0.5 μm, after which the displacement increases with decreasing particle size. These results suggest that the chances of deposition for large particles are higher than those for smaller particles and particles of 0.5 μm in diameter are the least likely to deposit in the airways.

3.2.4   Interception

Interception is the mechanism of deposition of a particle when the centre of gravity is within the streamlines of the gas phase but a distal end of the particle has touched a surface. Deposition by interception in the airways is important when the dimension of the airway radius is comparable to the dimensions of the particles. Thus, elongated particles are more likely to deposit by interception than spherical particles of similar volume (Timbrell, 1965; Johnson and Martonen, 1994). Since particles of therapeutic materials which require pulmonary deposition are very much smaller than the dimensions of the air space, interception should not be an important factor in well-defined formulations. Exceptions to this would be elongated particles, i.e. particles large in one dimension but with small aerodynamic diameters. The intentional interception of carrier particles in the upper airways could be advantageous in protecting the deep lung region from excessive exposure to these so-called inert particles.

3.2.5 Electrostatic precipitation

Significant electrostatic charges may be imparted to the droplets and particles of an aerosol at the generation stage (John, 1980). Drug particles are usually inhaled immediately after generation without the charge being neutralised and thus such charges will influence the overall deposition of the particles in the airways. A charged particle may induce an image charge on the surfaces of the airways and subsequently deposit by electrostatic precipitation (Chan and Yu, 1982). The repulsion between like-charged inhaled particles (space charge forces) may also direct the particles toward airway walls, resulting in deposition in those regions (Chen, 1978). However, the deposition of charged particles in the airways has not been extensively studied and still requires more investigation.

3.3 DELIVERY DEVICES

Medicinal aerosol delivery devices must generate an aerosol with most of the drug particles less than 10 μm in size, ideally in the range of 0.5–5 μm, in order to obtain a deep lung penetration after inhalation. However, the exact size range should be dependent upon the targeted sites of the medicaments. The device should also produce reproducible drug dosing and provide an environment where the drug can maintain its physico-chemical stability. It must also be easy to use, portable and inexpensive. Despite the numerous methods that can be employed to generate aerosols in therapeutically-useful size ranges and concentrations, only three basic aerosol delivery systems have found their way into commercial markets. They are metered dose inhalers (MDIs), nebulisers and dry powder inhalers (DPIs).

3.3.1 Metered dose inhalers (MDIs)

The introduction of MDIs in the 1950s was a significant innovation. Since then, they have emerged as the primary device for the outpatient treatment of asthma. Many medicaments have been introduced into this type of device, where the drug is contained in a pressurised canister either dissolved or dispersed in a suitable liquid propellant (or mixture of propellants). These propellants, mostly chlorofluorocarbons (CFC), are liquidified gases, and mixtures of these may be used to obtain optimal solution properties and desirable pressure, delivery and spray characteristics. MDIs may also contain suitable auxiliary materials such as solvents, solubilising agents, suspending agents and a lubricating agent to prevent clogging of the valve. The spray released from the MDI every time it is fired or 'actuated' consists of a metered volume of propellant containing a predetermined dose of drug.

On actuation of the MDIs, rapid initial vaporisation of the propellant occurs. This breaks up the liquid stream into filaments then droplets. The particles from MDIs usually have a high velocity which exceeds the inspiratory flow rate, thus a large number of particles impinge onto the oropharynx. A rapid cooling due to heat transfer during propellant evaporation promotes water vapour condensation onto the droplets. The resultant particle size might therefore be larger than the original micronised drug particles.

MDIs are small, portable and quick to use. They are able to deliver a precise unit dose and thus give a reproducible lung deposition. However, over the years a number of deficiencies have been identified. For example, only a small fraction of the drug escaping the inhaler penetrates the patient's lungs (Newman *et al.*, 1981) due to a combination of high particle exit velocity and poor co-ordination between actuation and inhalation. The unstable physical properties of the drug particles suspended in the propellants, combined with suboptimal valve designs, have led to reports of irreproducible dose metering following a period of storage (Cyr *et al.*, 1991). The propellants have also been found to extract low concentrations of potentially carcinogenic compounds from the valve components, which can subsequently be inhaled by patients (Sethi *et al.*, 1992).

The inability of many patients to synchronise aerosol actuation with inspiration is the problem of most concern in clinical application since many patients, even after extensive training, still cannot operate the device properly. To solve this problem, several inhalation aids have been developed. A spacer fitted to the mouthpiece of a conventional MDI is believed to cause a delay between actuation and inhalation thus reducing the need for co-ordination. The spray is actuated into the spacer device, which prolongs propellant evaporation, resulting in the formation of smaller droplets or particles and less reflex coughing and exhalation due to local cooling of the throat by impacted, evaporating droplets. A large portion of drug that would otherwise deposit into the oropharynx is retained in the spacer and this reduces systemic drug absorption and thus minimises local side-effects.

The second drawback is associated with the CFC propellants employed in MDIs. These propellants are considered environmentally harmful as they contribute to ozone depletion as a consequence of the chlorine atoms released by the effect of sunlight. In addition, CFCs are also potent greenhouse gases which contribute to global warming. With the commitment to phase out CFC propellants, investigation of alternative non-CFC propellants has been an active field and two are being developed, namely hydrofluoroalkene-134a (HFA-134a) and HFA-227. HFA-134a has been used in Airomir® (3M Healthcare), an MDI used to deliver salbutamol sulphate. However, replacement of CFCs with HFAs has not been an easy task, since these two kinds of propellants have different physico-chemical properties such as polarity, vapour pressure, density and so on

(MacDonald and Martin, 2000). These properties are crucial for the satisfactory formulation of MDIs. Although less than for CFCs, HFAs still contribute to global warming and they are also likely to cause environmental concern in the future.

3.3.2 Nebulisers

A nebuliser is a device for turning an aqueous solution of a drug into a mist of particles for inhalation (Horsley, 1988). They are mostly used in the treatment of patients who cannot use other devices successfully and for delivering large doses of bronchodilators and unusual medicaments that it is not possible to formulate into MDIs or as dry powder formulations. There has been a resurgence of interest in nebulisers in recent years due to their ability to generate small droplets capable of penetrating deeply into the lung, their high dose delivery capacity and miniaturisation of hardware. The development of high-output nebulisers has permitted shorter treatment times and increased efficiency of drug delivery to the patients (Clay *et al.*, 1987; Johnson, 1989). Nebulisers may also represent ideal delivery systems for biotechnologically-derived proteins and peptides. There are two common types of nebuliser, air-jet and ultrasonic, with many commercially available examples of each type.

Air-jet nebulisers operate by forcing a compressed gas such as air or oxygen through a narrow bore inlet and across the top of a capillary tube, the lower end of which is immersed in the liquid to be nebulised. On leaving the inlet, the gas expands resulting in the generation of a negative pressure which draws the liquid up the capillary and into the gas stream. This breaks the liquid into droplets, the largest of which are removed by baffles and returned to the liquid and the smaller droplets are delivered to the patient through a face mask or mouthpiece.

Ultrasonic nebulisers, however, do not require the use of a carrier gas. The respiratory solution or suspension is atomised by means of a piezoelectric crystal transducer. An alternating current causes the shape of the crystal to alternately shrink and expand, causing a vibration that is amplified by a stainless steel shim. This vibration is then transferred to the solution or suspension in the nebuliser reservoir, leading to its fragmentation.

The prime advantages of nebulisers are their simplicity, generation of relatively small droplets is easy, and doses as high as 1 g can be delivered. However, they are not historically portable devices and have traditionally been limited to the treatment of hospitalised or non-ambulatory patients. They are also expensive when compared to other inhalation devices, and the treatment is time-consuming.

3.3.3 Dry powder inhalers

One of the consequences of the therapeutic use of penicillin in the late 1940s involved a variety of attempts to develop devices which would apply 'dusts' to the respiratory tract for local therapy (Bell, 1992). However, not until the 1960s, was the first successful dry powder inhaler, the Spinhaler®, introduced for the delivery of disodium cromoglycate (Bell *et al.*, 1971). Over recent years, dry powder inhalers (DPIs) have become more and more attractive since they possess many advantages over MDIs and nebulisers. Without the use of any volatile propellants, dispersion and entrainment of drug particles in DPIs are caused by the inhalation efforts of the patient alone. Thus, DPIs are environmentally friendly and very easy to use and handle, overcoming the problems associated with the required synchronisation of actuation and inhalation often required to operate MDIs successfully. DPIs are highly portable and relatively inexpensive in comparison to nebulisers. Since the drugs are also kept in the solid state in DPIs it is expected that they would exhibit a high physico-chemical stability. This is of particular relevance when considering the formulation of proteins or peptides for pulmonary delivery, for example.

In most DPIs, drug particles are micronised and adhered to inert carrier particles. A number of different carriers have been used although lactose has been employed the most frequently, only because it has a history as a widely used and safe excipient in solid dosage forms (Timsina *et al.*, 1994). The carrier particles are designed to be of such a size that after inhalation, most of them remain in the inhaler or deposit in the mouth or upper airways. In order to reach the lower airways, the drug particles must therefore dissociate from the carrier particles and become dispersed in the air flow. The redispersed drug particles may then undergo inertial impaction in the mouth, on the back of the throat and the upper airways. Some particles (2–5 µm) are likely to reach and deposit in the lower airways through gravitational sedimentation, interception and Brownian diffusion. Therefore, three major processes are involved in the delivery of drugs from DPIs, namely, the detachment of drug particles from the carrier, dispersion of drug particles in the air flow and deposition in the respiratory tract. Any factor that affects any of these processes could ultimately influence the deposition properties of the inhaled particles. As far as the dry powder inhalation system is concerned, the dominant factors are the design of the inhaler device and the powder formulation: very often it is necessary to consider these two aspects together.

3.3.3.1 Design of inhaler devices

The design and engineering of dry powder inhalers (DPIs) is still an evolving science. An ideal DPI should be simple to use, compact and inexpensive, produce high aerosolisation efficiency with low inspiratory effort, produce reproducible emitted and fine particle doses, preferably be a multidose device and maintain physico-chemical stability of the powder formulations (Timsina *et al.*, 1994). Many types of DPIs have been made commercially available (Figure 3.2) but none of the currently available DPIs meet all the requirements.

The Spinhaler® employed a gelatin capsule to contain the metered dose of the micronised drug particles that had been mixed with lactose carrier particles (Bell *et al.*, 1971).

The capsule is placed in a rotor and before inhalation, the drug particles are released by simply actuating a device which pierces the capsule with metal needles. During inhalation, the capsule revolves with the rotor and the powder disperses through the holes created in the sides of the capsule walls into a relatively wide air channel. In order to generate effective dispersion of the fluidised powder mixture, an air flow rate of at least 35 l min^{-1} is needed. Another similar device for the delivery of sodium cromoglycate has been introduced whereby the gelatin capsule is pierced at both ends by means of small needles. During inhalation, the capsule rotates in a small chamber and the powder disperses through the holes into the inspired air. In contrast, the stationary capsule principle was employed to deliver a bronchodilator – fenoterol (Berotec®, Boehringer Ingelheim).

In 1977, the Rotahaler® (Allen and Hanbury) was launched for the delivery of salbutamol and beclomethasone dipropionate. In this device, gelatin capsules are broken open before inhalation, by twisting the segments of the device. The broken capsule shell then drops into a relatively large chamber and drug powder is liberated for inhalation.

All the early devices described above are single unit or 'single shot' devices which require reloading every time before use. This is problematic especially when the patient is having a severe asthmatic attack. This prompted the development of multidose devices, the first of which was introduced by Astra Pharmaceuticals, the Turbohaler®. In

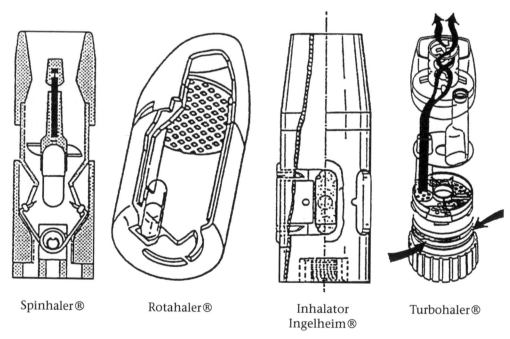

Spinhaler®   Rotahaler®   Inhalator   Turbohaler®
                                           Ingelheim®

Figure 3.2: Illustrations of some of the most commonly used DPIs. (Adapted from Timsina *et al.*, 1994.)

this device, a disposable reservoir is employed to meter the drug on priming the inhaler. The dosing unit consists of a disc with a group of conical holes and the powder is pushed into them by scrapers upon rotation of the dosing unit. Up to 200 doses can be delivered by this device. In the late 1980s, another multidose inhaler, the 'Diskhaler', was launched by Glaxo. Each metered dose of the drug is contained within one of either four or eight blisters in a circular aluminum foil disc, which is inserted into the device. The blisters are pierced in turn to free the powder, mixtures of micronised drug particles and inert carrier particles such as lactose, for inhalation. The tray must be slid out and in to prepare the device for the next dosing. The Diskhaler® was originally developed for the delivery of salbutamol and beclomethasone dipropionate but more recently it has also been used for salmeterol and fluticasone propionate.

Other DPIs include the Ultrahaler® (Fisons) and Diskus® (GlaxoWellcome). The Ultra-haler® is very similar to the Turbohaler® but compacted drugs are used and it is claimed to be of use for the administration of high-dose compounds. The Diskus® combines the benefits of pre-measured, individually sealed and protected doses ensuring protection from the environment in a multi-dose device (Fuller, 1995).

3.3.3.2 Peak inspiratory flow rate (PIFR) via different DPIs

The flow rate at which a patient inhales through a dry powder inhaler significantly affects the amount of drug reaching the lung since most of the currently available dry powder inhalers rely on the patients' inspiratory flow rate for the delivery and aerosolisation of the medicament. It is the patient's inspiratory effort that causes the air flow, which in turn imparts energy to the formulation. For example, the total amount of drug delivered via a Turbohaler® after inhalation at a flow rate of $60 \, l \, min^{-1}$ was approximately twice as much as that delivered at $28.3 \, l \, min^{-1}$ (Schultz et al., 1992). The variability in the delivered dose was also found to decrease at flow rates higher than $28.3 \, l \, min^{-1}$. An air flow rate of at least $35 \, l \, min^{-1}$ was needed to generate effective dispersion of the fluidised powder mixture into the inhaled air through a Spinhaler® (Bell et al., 1971).

Different DPIs have different design features. Such a difference may have at least two consequences. First, different inhaler devices may produce unique flow rates after inhalation at similar inspiratory efforts. Second, drug dispersion and deaggregation may follow a different pattern after inhalation at a similar flow rate from the various inhaler devices.

The Peak Inspiratory Flow Rate (PIFR) attained by asthmatic patients is reported to be on average of $150 \, l \, min^{-1}$ (Brown et al., 1995). During submaximal inhalation through the Diskhaler®, the PIFR was found to range from $59–170 \, l \, min^{-1}$ (Spiro et al., 1992) whilst the unimpeded PIFR can easily be in excess of $200 \, l \, min^{-1}$ (Timsina et al., 1993). A reduction in the PIFR is observed in the presence of a DPI, the reduction being device dependent (see Table 3.2).

TABLE 3.2

Peak inspiratory flow rate (PIFR) after inhalation from different DPIs by healthy male and female volunteers (Timsina *et al.*, 1993).

| Device | Peak inspiratory flow rate (PIFR) ($l\,min^{-1}$) | | | |
| | Male | | Female | |
	Mean	SD	Mean	SD
Control	333.0	97.6	213.9	70.5
Rotahaler®	216.6	45.6	160.1	43.7
Spinhaler®	185.7	40.8	133.6	36.0
Turbohaler®	81.8	12.8	55.6	13.0
Inhalator Ingelheim®	70.4	18.6	48.3	13.2

As can be seen, the Rotahaler® is a less resistive device when compared to the Turbohaler® or Inhalator Ingelheim® in both male and female volunteers. Inspiratory flow rates above $60\,l\,min^{-1}$ are required for efficient actuation of the Rotahaler®. However, in order to achieve $60\,l\,min^{-1}$ through a Turbohaler® nearly three times the effort needed for the Diskhaler® would be required (Sumby *et al.*, 1992). It has been calculated that at maximum inspiration, the amount of work performed through a high resistance inhaler is approximately 70% higher than that through a low resistance device (Deboer *et al.*, 1996). The Turbohaler® was shown to be an effective delivery system, at inspiratory flow rates around $30\,l\,min^{-1}$, when, although drug delivery to the pulmonary region was reduced, this was not mimicked by the clinical response (Dolovich *et al.*, 1988; Newman *et al.*, 1991). It has also been shown that instructing patients to take a 'forceful and deep' inhalation optimised the use of the Turbohaler® and increased PIF by 20% to $40\,l\,min^{-1}$ (Persson *et al.*, 1997).

3.3.3.3 Device resistance

The resistance to the inhalation air stream is a critical parameter to be considered in terms of the design of the inhaler device. Two opposing factors are involved in choosing a suitable resistance of an inhaler device. On the one hand, increasing the device resistance generally increases the turbulence of the air stream, which in turn increases the dispersion and deaggregation of aerosol dry powders. On the other hand, an increased device resistance would require a higher inspiratory effort by the patient in order to achieve a suitable air flow rate and this may prove to be problematic for severely asthmatic patients, children and infants. Therefore, these two opposing factors have to be balanced when introducing resistance to the inhaler device.

Device specific resistance (R_D), in $(cmH_2O)^{0.5}\,l^{-1}\,min$, is the main factor which controls the operational flow rate of a powder inhaler (Clark and Hollingworth, 1993). Assuming that a device exhibits frictionless airflow and that the pressure fall is sufficiently small so as to be able to neglect changes in air density, the volumetric flow through a device would be expected to be proportional to the square root of the pressure drop developed across it (Ower and Pankhurst, 1977). Then, R_D is given by the slope of the plot of volumetric flow against the square root of the pressure drop.

A general equation to describe the relationship between flow rate and pressure drop is given as follows (Olsson and Asking, 1994):

$$\Delta P = R_D Q^X \tag{3.9}$$

Where ΔP is the pressure drop in cmH_2O, R_D is the device resistance, Q is the volumetric flow rate in $l\,min^{-1}$ and X is a dimensionless constant.

The device resistance can then be calculated from the intercept using the logarithmic transformation of the above equation:

$$\log \Delta P = \log R_D + x \log Q \tag{3.10}$$

This would result in a linear relationship between $\log \Delta P$ and $\log Q$.

Table 3.3 shows the air flow resistance for some commercially available dry powder inhalers.

As can be seen, all the inhaler devices impart some amount of resistance to the inhaled air stream as compared to normal airway resistance values in adults, which range from 0.012 to 0.025 $(cmH_2O)^{0.5}\,l^{-1}\,min$ (Cole and Mackay, 1990). The Rotahaler® was the least resistive device tested whilst the Inhalator Ingelheim® was found to be the most resistive to the inhaled air stream. The variation in the resistance values can explain the different PIFRs, as outlined in Table 3.2, after inhalation from these devices. For example, the Rotahaler® produced the highest PIFR due to it having the lowest R_D of all the devices and the reverse trend is true for the Inhalator Ingelheim®.

DPIs with lower resistance usually require higher inspiratory flow rates than those with higher air flow resistance to achieve similar delivery efficiency of drugs. For example, the Turbohaler® operates well at flow rates as low as 33 $l\,min^{-1}$ (Persson et al.,

TABLE 3.3

Airflow resistance for some commercially available DPIs

Dry powder inhalers	Resistance to airflow $(cmH_2O)\,l^{-1}\,min$
Rotahaler®	0.040
Spinhaler®	0.051
Diskhaler®	0.067
Turbohaler®	0.100
Inhalator Ingelheim®	0.180

1988) whilst the Rotahaler® generally requires much higher flow rates for powder inhalation (Bogaard et al., 1989). Dry powder devices which operate at low inspiratory flow rates, e.g., the Diskhaler®, and the Turbohaler®, are clinically desirable for children and adults with decreased lung function either because of age or disease. The Inhalator Ingelheim® also provides effective bronchodilation at 16 to 19 l min^{-1} (Pedersen et al., 1986) compared to the ISF® (Cyclohaler®) whose optimal performance is at flow rates of 40 to 80 l min^{-1} (Zanen et al., 1992). The difference in the flow characteristics generated in the various dry powder inhalers will result in different delivery efficiencies of drug particles. For example, different lung deposition profiles of radiolabelled sodium cromoglycate were obtained after delivery of the drug particles from four different inhaler devices, namely, Spinhaler®, Cyclohaler®, Berotec® and Rotahaler® (Adjei and Gupta, 1997). The Cyclohaler® and Berotec® were most effective in delivering the drug to the lung producing peripheral deposition of 15% of the dose. The Spinhaler® was less effective with peripheral deposition of about 10% of the dose. The Rotahaler® was found to be the least effective with only 5% of the drug being delivered to the peripheral regions of the lung. However, powder delivery devices are usually designed to deliver a specific product to the lung and its performance criteria are based on that drug or formulation. For example, the Rotahaler® was designed to deliver salbutamol and beclomethasone not sodium cromoglycate. Its relatively poor performance in delivering sodium cromoglycate does not necessarily indicate a poor performance of this device for other drugs. Therefore, the factors separating functional performance of formulations should be assessed independently of device configuration before any attempts to draw meaningful conclusions from biological equivalence results with various formulations of the same drug.

3.4 PHYSIOLOGY OF THE HUMAN RESPIRATORY TRACT

The respiratory tract is a specialised and unique region of the body. It is responsible for the gaseous exchange between the atmosphere, systemic circulation and the cells of the body, the homeostatic maintenance of blood pH and the provision of vocal expression. The human respiratory tract can be divided into the upper and lower respiratory regions. The upper being the nose, nasal passages, paranasal passages, mouth, eustachian tubes, pharynx, larynx, trachea and bronchi. The lower respiratory tract consists of the bronchioles and the alveoli (Figure 3.3). Thus, the respiratory tract can be viewed as a series of branching passageways, originating at the trachea and terminating at the alveolar sac. The branching tree starts off asymmetrically when at the bottom of the trachea, the right main bronchus continues sharply downwards, whilst the left main bronchus branches sharply to the left. Three lobes are found on the right side of the chest and two on the left. The branching within the lobes is dichotomous, with a mean branching angle of 37° from the straight-on position.

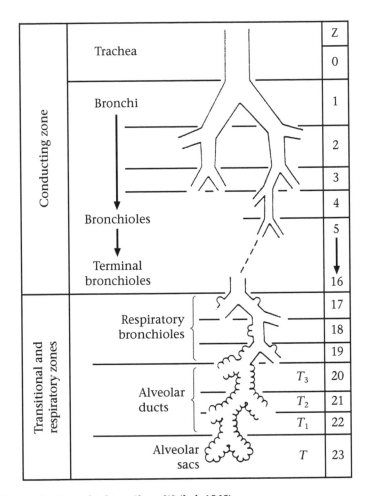

Figure 3.3: The tracheobronchial tree (from Weibel, 1963).

The lower airways can be further divided into two zones: the conducting zone and the respiratory zone. The conducting zone, akin to a canal leading to the central regions of the lung, provides humidification, gas buffering, gas transference and air warming. The conducting airways are surrounded by smooth muscles which contribute to the increase and decrease in the tracheal diameter during inspiration and expiration. They are also lined with specialised cells, some of which secrete mucus whilst others bear cilia, the co-ordinated beating of which sweeps the mucus upwards, an effect often called the mucociliary escalator, and by doing so it promotes homeostasis. Succeeding the conducting zone is the respiratory zone and this is the site where gas exchange takes place.

Based on experimental data, a morphological model of the human lung has been proposed by Weibel (1963) and has become widely accepted in the investigation of particle deposition. This model is a symmetrical, dichotomously branching network of cylindrical tubes, (Figure 3.3) and from the trachea, through the terminal bronchioles, to the alveolar sacs, the airways are divided into 24 generations. Each airway divides to form two smaller daughter airways and thereby the number of airways at each generation is double that of the previous one; there are 2^I (where I is the number of generations) airways in each generation. The ICRP Task Group on Lung Dynamics (1966) proposed that the respiratory tract be divided into three compartments, namely, nasopharynx, tracheobronchial and pulmonary regions. As most dry powders are administered via oral inhalation, it is more relevant to consider the oropharynx region rather than the nasopharynx region. The tracheobronchial region was defined as the non-alveolated bronchial airways including the trachea and terminal bronchioles (generations 0–16), its function being mainly air-conduction. The pulmonary region was the gas-exchange compartment inclusive of respiratory bronchioles and alveolar ducts, corresponding to the generations 17 to 23 of the model (Weibel, 1963).

In passing from the mouth through the trachea to the alveolar sac, several physiological changes occur in the airways that are important in determining particle deposition (Table 3.4). First, the airway diameter decreases with increasing generations whereas the number of airways for each generation increases at a much higher rate (double the previous generation). Thus, whilst the calibre for the individual airway decreases, the total area for that generation increases drastically. This change has a profound impact on the air-flow dynamics and eventually influences the deposition of particles. Second, the upper respiratory tract (all airways distal to the trachea) is composed of the mouth, pharynx and larynx. Whilst moving through this region, the air stream is subject to a sharp change in direction especially from pharynx to larynx (which is a short cavity containing a slit-like opening in its central portion). This structure apparently induces instability of the air stream (Pedley et al., 1977) and increases the chances of deposition by impaction. Third, the tracheobronchial region contains large numbers of ciliated cells, which sweep a sheet of mucus towards the pharynx, from which it is swallowed. Thus, particles deposited in this region will be rapidly removed. Deposition in the region is not uniform with most drug particles being preferentially deposited at the bifurcations and on the bulged cartilagenous rings (Martonen, 1993). In the lower airways, the air stream becomes more stable (Sudlow et al., 1971) and the majority of the remaining airborne particles will deposit mainly by sedimentation and diffusion mechanisms.

Thus, large particles, such as those >10 μm, will most probably impact in the mouth and on the back of the throat (Table 3.4), and these will eventually be swallowed and be delivered to the gastrointestinal tract. The pharynx, larynx and the entrance to trachea provide a major region for deposition of airborne particles, especially by impaction. The transfer through the larynx provides significant resistance to airflow and induces considerable turbulence downstream. This flow disturbance may persist for several generations

TABLE 3.4

Some characteristics of the respiratory tract related to drug deposition in different pulmonary regions. Calculations assume a steady inspiratory flow of 60 l min^{-1}: D_{sed} is the distance that a 1-μm particle will sediment during transit to that region; Sk is the Stokes' number for a 1-μm particle in that region; R_e is the Reynolds' number for that region (adapted from Davies, 1961).

Region	Number	Diameter (cm)	Length (cm)	Residence time (s)	D_{sed} ($\times 10^{-4}$ cm)	Sk ($\times 10^{-3}$)	R_e
Mouth	1	2	7	0.022	0.7	0.48	
Pharynx	1	3	3	0.021	0.7	0.14	
Trachea	1	1.7	12	0.027	0.9	0.78	5010
Main bronchi	2	1.3	3.7	0.010	0.3	0.88	3300
Lobar bronchi	5	0.8	2.8	0.007	0.2	1.50	2220
Segmental bronchi	18	0.5	6	0.021	0.7	1.71	942
Intrasegmental bronchi	252	0.3	2.5	0.045	1.5	0.57	225
Bronchioles	504	0.2	2.0	0.032	1.1	0.95	84
Secondary bronchioles	3024	0.1	1.5	0.036	1.2	1.27	28
Terminal bronchioles	12,100	0.07	0.5	0.023	0.7	0.93	10
Respiratory bronchioles	1.7×10^5	0.05	0.2	0.067	2.2	0.18	1
Alveolar ducts	8.5×10^5	0.08	0.1	0.44	14.5	0.01	0
Atria	4.2×10^6	0.06	0.06	0.71	23.4	0.01	0
Alveolar sacs	2.1×10^7	0.03	0.05	0.75	24.8	0.01	0
Alveoli	5.3×10^8	0.015	0.015	4	132	–	–

■ CHAPTER 3 ■

of the bronchial airways before becoming attenuated. A portion of remaining airborne particles can be expected to deposit in this region by impaction. Deposition in this region is often considered undesirable since it is rarely the target site for many drugs and deposited particles will, as has been stated, eventually be removed by the mucociliary escalator and swallowed.

In the lower tracheobronchial region ($I > 6$), the air stream gradually exhibits laminar flow. The Stokes number of the remaining airborne particles and the flow rate of the stream will gradually decrease such that deposition by impaction becomes less significant whilst deposition by sedimentation becomes more and more important. However, deposition at airway bifurcations still occurs preferentially. Deposition in this region (generations $6 < I < 16$) which includes the terminal bronchioles may not be desirable for all drugs, since drug particles will be removed by mucociliary action. However, this region includes the major sites of action for anti-asthmatic drugs.

In the pulmonary region, gaseous exchange between the bloodstream and the air takes place. Air flow is very slow, even non-existent, relying primarily on gaseous diffusion. Thus, the Stokes number of the particles that are still airborne becomes negligibly small, indicating minimal deposition by impaction. Due to the slow air flow, the residence time of particles in this region is greatly increased and hence deposition occurs mainly by time-dependent mechanisms such as sedimentation and Brownian diffusion. The pulmonary region is considered to be the ideal site of drug deposition especially for drugs administered to elicit systemic effects. Consequently, this region is the major target site for the delivery of such drugs.

3.5 BREATHING PATTERNS

The most important features of inhalation are the flow rate, the inhaled volume and the breath-holding time after inhalation (Newman *et al.*, 1982).

3.5.1 Flow rate

The inhalation flow rate can change the overall deposition of airborne particles by affecting their deaggregation from the formulation, their Stokes number and the turbulence of the inhaled air stream and, finally, their residence time in each section of the respiratory tract.

As discussed previously, increasing inhalation flow rates usually increases the deaggregation of the formulation, resulting in an improved fraction of small particles. However, in accordance with equation (3.2), an increased flow rate will produce a higher Stokes number for the airborne particles, which will in turn increase deposition by impaction, especially in the upper airways. Furthermore, increasing flow rates should enhance the turbulence of the air stream, mainly in the upper airways. This will result in a greater contact of particles with the airway walls, thereby increasing the deposition in this region, concomitantly decreasing the deposition in the lower airways. Increasing the flow rate will also reduce the residence time of airborne particles in the respiratory tract and thereby reduce the deposition of particles by sedimentation and diffusion mechanisms in the lower airways. Therefore, slow inhalation is desirable to obtain improved deposition of airborne particles in the lower airways. This has been demonstrated by many studies on deposition from MDIs and nebulisers (Newman *et al.*, 1981 and 1982). Slower inhalation usually results in higher deposition in the lower airways since these devices do not need the inhaled air stream to atomise the drug particles. For example, a slow inhalation of $30\,l\,min^{-1}$ was found to be preferential for the delivery of a bronchodilator in terms of both pharmacological effects and deposition pattern than a fast inhalation at $80\,l\,min^{-1}$ (Newman *et al.*, 1981, 1982). Faster inhalation was less effective since more aerosol impacted in the oropharynx and was unable to penetrate into the lung. By means of mathematical modelling (Martonen and Katz, 1993), the total lung deposition was shown to be inversely related to the inspiratory flow rates due to increased inertial impaction at higher flow rates. Pulmonary deposition increases with decreasing flow rate due to prolonged residence times. Thus, the tracheobronchial deposition of 6-μm particles was markedly increased from 30% at an inhalation flow rate of $0.5\,1\,s^{-1}$ to 50% at $0.04\,1\,s^{-1}$ and the deposition in the smaller ciliated airways was also increased (Anderson *et al.*, 1995). This phenomenon was attributed to less deposition in the mouth and throat due to a reduced impaction at slower inhalation flow rates.

The influence of flow rate on deposition is more complicated for dry powder inhalers when compared with MDIs and nebulisers. The breath-actuated nature of most dry powder inhalers requires that sufficient energy be generated since too slow a flow rate may not be sufficient to detach the adhesive particles from their carriers and deaggregate cohesive drug particles. Therefore, the total amount of drug particles reaching the lower airways may be much lower at slower inhalation rates than at faster inhalation rates, although the probability of the deaggregated individual particles reaching the lower airways may be higher in the former case. Thus, unlike MDIs and nebulisers, DPIs may require a higher inhalation flow rate to effectively deaggregate the drug particles. In a study with fenoterol powder, Pedersen and Steffensen (1986) demonstrated that a significantly increased response was observed in children who inhaled as fast as possible, in comparison to children who inhaled only slowly at 16–19 l min^{-1}. Auty *et al.* (1987) found that for sodium cromoglycate (SCG) delivered from a Spinhaler® a higher plasma concentration of drug was achieved at higher inspiratory flow rates and that a peak inspiratory flow rate around 160 l min^{-1} probably represented the optimal inhalation rate. Pitcairn *et al.* (1994) used lactose and ^{99m}Tc salbutamol blends to investigate the influence of flow rate on deposition of the drug particles from a multidose Pulvinal® inhaler. A significantly higher percentage of the nominal dose ($p = 0.05$) was observed to deposit in the lungs at a flow rate of 46.0 l min^{-1} (14.1 ± 3.2%) as compared to that (11.7 ± 2.3%) attained at the slower flow rate of 27.8 l min^{-1}. Newman *et al.* (1994) investigated the deposition of ^{99m}T$_c$ labelled sodium cromoglycate powder from a Spinhaler® at fast (120 l min^{-1}) and slow (60 l min^{-1}) peak inhaled flow rates. Inhalation at 60 l min^{-1} significantly ($p < 0.01$) reduced deposition in the lungs compared to inhalation at 120 l min^{-1}. However, Zanen *et al.* (1992) showed that there was no significant difference in bronchodilation by salbutamol after inhalation at 40 or 80 l min^{-1} through a Cyclohaler® as particles <5 μm were reported to be separated completely at the lower flow rate.

3.5.2 Tidal volume and breath-holding

It has long been demonstrated that deep inhalation and breath-holding can improve the deposition of an aerosol in the lung. For example, Riley *et al.* (1976, 1979) found that an isoprenaline MDI was more effective when actuated at a high lung volume, i.e. 80%

vital capacity (VC) compared with a low lung volume, 20% VC. A period of breath-holding after inhalation increases the probability of peripheral deposition since it allows time for the particles to settle on the airways by gravitational sedimentation and diffusion. Newman *et al.* (1982) demonstrated that the deposition of ^{99m}Tc teflon particles (aerodynamic diameter 3.2 μm) in patients with obstructive airways disease was greatly influenced by the breath-holding time. At a given vital capacity of 20, 50 or 80%, the increase in breath-holding from 4 to 10 s consistently improved the deposition of teflon particles in the whole lung, with the increase ranging from 6.0 to 14.3% of the administered dose. However, increasing vital capacity failed to show the predicted commensurate increase in the whole lung deposition. Auty *et al.* (1987) assessed the blood levels of sodium cromoglycate following inhalation from the Spinhaler® and concluded that an optimised lung deposition was obtained by a combination of fast inhalation and subsequent breath-holding. More recently, Martonen and Katz (1993) employed a mathematical model to predict the effects of tidal volume and breath-holding on the deposition of an inhaled aerosol. According to their calculations, increasing the tidal volume should increase the whole lung deposition and pulmonary deposition for all particle sizes. This increased deposition was attributed to the deeper particle penetration into the lung at the increased tidal volumes and the increased particle residence time in the peripheral airways, leading to improved deposition due to sedimentation and diffusion.

3.6 PHYSICO-CHEMICAL PROPERTIES OF DRUG PARTICLES

Apart from inhalation manoeuvres, formulation factors are also important in determining the deposition of inhaled particles in the respiratory tract. Such factors include particle size, size distribution, density, shape and hygroscopicity.

3.6.1 Particle size

The deposition characteristics and efficacy of an aerosol depend largely on the particle or droplet size (Newman and Clarke, 1983; Moren *et al.*, 1985; Heyder *et al.*, 1986). Inhalation aerosols may be targeted to a particular deposition site by governing the particle size. Large droplets of significant mass may achieve high velocity as a result of the drag forces of inhalation. Their momentum will render them more likely to deposit in the upper airways through inertial impaction. Small particles may pass through the upper airways and deposit in the lower airways under the influence of gravity and Brownian motion. However, too small a particle may hang in the air stream and, consequently, be exhaled.

According to equations (3.2) and (3.5), the impaction and sedimentation are proportional to the product of density and the square of the diameter (ρd^2) which can be used to predict the deposition site. For example, the empirical deposition probability formula describing extrathoracic deposition of a particle after mouth breathing was given as (Martonen, 1993):

$$P(m) = 0, \text{ when} \qquad \rho d^2 V < 1.67 \times 10^3$$
$$P(m) = -0.496 + 0.154 \log (\rho d^2 V), \text{ when} \quad 1.67 \times 10^3 < \rho d^2 V < 10^4$$
$$P(m) = -2.988 + 0.777 \log (\rho d^2 V), \text{ when} \quad 10^4 < \rho d^2 V < 8.3 \times 10^4 \qquad (3.11)$$

where ρ is the particle density in $g\,cm^{-3}$, d is the particle diameter in μm and V is the mean inspiratory flow rate in $cm^3\,s^{-1}$.

Thus, for a given inspiratory flow rate V, the higher the value of ρd^2, the more likely the particle is to deposit in the extrathoracic compartment and less likely to reach the lower airways. Therefore, the square route of this product (ρd^2) is known as the aerodynamic diameter of the particle, i.e. the diameter of a unit density sphere with the same settling velocity as the particle in question (Agnew, 1984).

The influence of particle size on deposition has been the subject of many reports and numerous mathematical models have been proposed for calculating the extent of retention of particulate material of various sizes in the average adult human lungs (Martonen, 1993). For example, the Task Group on Lung Dynamics, which concerned itself mainly with the hazards of inhalation of environmental pollutants, proposed a model for deposition and clearance of particles from the lung after nose-breathing (Figure 3.4)

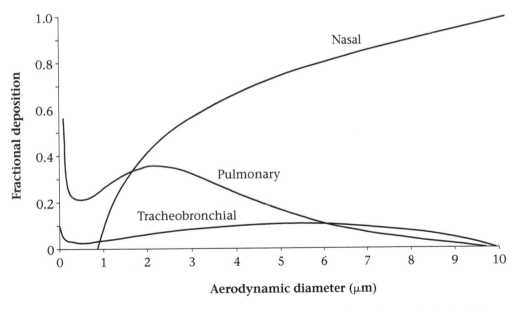

Figure 3.4: Effects of particle size on the deposition of aerosols in the respiratory tract (Task Group on Lung Dynamics, 1966).

(Task Group on Lung Dynamics, 1966). The model suggested that particles larger than 10 μm in diameter were most likely to deposit in the mouth and throat. Between the sizes of 5 and 10 μm, deposition from mouth to airways would occur. Particles smaller than 5 μm in diameter were predicted to deposit more frequently in the lower airways. Thus, pharmaceutical powders would require a particle size less than 5 μm in order to reach the lower airways.

Davies *et al.* (1976) estimated the ideal size for therapeutic aerosols to be between 0.5–7.0 μm. Particles larger than 7.0 μm are not believed to be able to penetrate beyond the trachea, whereas particles smaller than 0.5 μm are most probably exhaled. Curry *et al.* (1975) found a greater response to sodium cromoglycate when administered at a size of 2.0 μm compared with particle sizes of 6.0 and 11.7 μm. This observation was confirmed by Rees *et al.* (1982), who reported greater bronchodilator effects with terbutaline sulphate of diameter less than 5.0 μm compared with the particles of 5–10 μm or 10–15 μm. An improved deposition in the lower airways of smaller radioactively labelled teflon particles in comparison to larger sized particles has also been observed (Rees *et al.*, 1982).

Generally, it is widely accepted (Gonda, 1992) that particles with aerodynamic diameters greater than about 15 μm will deposit entirely in the extrathoracic regions, whereas particles with aerodynamic diameters of 5–10 μm will deposit in the tracheobronchial regions. Particles with diameters of 1 to 3 μm are most likely to deposit in the lower airways, with a maximum alveolar deposition of about 60%. However, even at this optimal alveolar deposition, a significant amount of material will still deposit in extrathoracic regions (10%) and tracheobronchial regions. Under normal breathing, submicron particles are exhaled, especially particles with diameters of about 0.5 μm, since they do not have enough mass and momentum to deposit by impaction and sedimentation. At the same time, they are not small enough to exhibit significant Brownian motion and consequently they fail to be removed from the air stream due to diffusion mechanisms. However, ultrafine particles (<0.01 μm) may deposit in the lower airways sufficiently rapidly through Brownian motion but the use of ultrafine particles is limited, since these are difficult to generate in high concentrations and carry much less mass than larger particles. For an aerosol to be effective, the fraction deposited at the targeted sites must be equal to or exceed the therapeutic dose. Therefore, large particles that are able to penetrate into the deep lung offer the greatest therapeutic advantage. Hence, the target size of inhaled particles is generally accepted as 1 to 5 μm (Smith *et al.*, 1980) and this size range has been used to guide aerosol generation technology.

3.6.2 Particle size distribution

The influence of size distribution on the deposition of inhaled aerosols is related to the particle size of the individual particles. Most aerosols are polydisperse and have a wide range of diameters. The particle size distribution of an aerosol produced by an MDI, DPI and nebuliser usually follows a log normal distribution, that is to say the fraction of particles of a particular size when plotted against the logarithm of the particle diameter, exhibits a normal, bell-shaped or Gaussian distribution (Raabe, 1978).

The conventional measures used to express the particle size and distribution are the median diameter and geometric standard deviation (GSD). The median diameter of an aerosol is the value of the diameter corresponding to 50% on a plot of diameter versus cumulative fraction such that 50% of all particles are above and below this value. The median diameter can be expressed either as a geometric median diameter (GMD), which is characterised by the number of particles, or a mass median diameter (MMD), which is based on the mass or volume of the particles. The size range is expressed in terms of the GSD which is the ratio of the diameters of the cumulative fractions between 50% and 84.1% or 50% and 15.9%, the value corresponding to the standard deviation of the median diameter of the curve (see Chapter 7).

3.6.3 Particle density

The aerodynamic size of a particle is a function of the square root of its density and, hence, for a particle of a given geometric particle size, reducing its density would decrease its aerodynamic particle size. It is this simple theory that led to the utilization of large porous particles for pulmonary drug delivery (Edwards *et al.*, 1997). It was found that inhalation of large, porous insulin particles resulted in higher systemic levels of the drug as compared with that achieved by the inhalation of small, non porous particles. This is discussed further in Section 5.5.7.1.

Particles with a narrow distribution all have a similar particle size, whilst particles with a broad distribution may have the same median diameter but possess a wide range of particle sizes. The log normal distribution of most aerosols indicates that there is often a long tail of particles with large aerodynamic diameters. These large particles contain a significant fraction of the administered dose, and it is the amount of drug that dictates the resultant pharmacological effect rather than the total number of drug particles. Therefore, particles of the same MMD but with different polydispersities will exhibit different deposition profiles in the lung and it is always desirable to produce aerosols with the lowest polydispersity. Several theoretical calculations have shown that the polydispersity of an aerosol could affect the regional deposition on inhalation. For example, Gonda (1981) calculated that there was a reduction in alveolar deposition from about 60% of the inhaled dose for a monodisperse aerosol with an MMD of 2 μm and a GSD (σ_g) of 1.0 to less than 30% for a polydisperse aerosol of the same MMD with a GSD of 3.5 (Figure 3.5).

CHAPTER 3

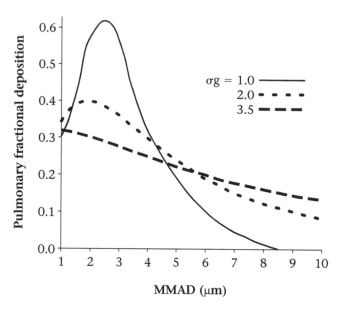

Figure 3.5: Effects of particle size distribution on aerosol deposition (after Gonda, 1981).

3.6.4 Particle shape

Particle shape has an important influence on the rheological properties of powders. Anisometric particles, which have an elongated or flattened shape, tend to build up open packings of higher porosity. Such particles tend to align along their longest axis in the direction of flow and thus exhibit less internal friction than more isometric particles (Neumann, 1967). In the field of aerosol science, the use of elongated particles has attracted much interest. Long objects, such as fibres and needle-like crystals, have aerodynamic diameters almost independent of their length and the diameter is approximately equal to the shortest dimension of the particle in question (Hinds, 1982). Thus, elongated particles may exhibit a much smaller aerodynamic diameter than spherical particles of similar mass or volume. Therefore, the controlled growth of crystals in one direction, or needle-like preparations, produced by other techniques often have a narrow distribution of aerodynamic diameters as long as their short axes are reasonably uniform, although they may have a wide range of lengths. As the deposition of particles is dependent on their particle size and distribution, elongated particles that have an aerodynamic diameter capable of penetrating into the deep lung will have higher mass than spherical particles of similar diameters. Furthermore, the dimension of the shortest axis of an elongated particle can be expected to be reasonably uniform. A narrow size distribution will confer upon the particles a higher selectivity of deposition to the different regions in the respiratory tract. Elongated particles which reach the alveolar region can deposit there by the mechanism of interception whereas spherical particles of similar aerodynamic diameter (<2 μm) will have a high probability of being exhaled unless prolonged breath-holding is exercised. Chan and Gonda (1989) were able to prepare elongated crystals of cromoglycic acid by precipitation of the drug with hydrochloric acid from aqueous solutions of cromolyn sodium and subsequent recrystallisation from hot water or mixtures of dimethyl sulphoxide and water. The aerodynamic size distribution, as measured in a cascade impactor, was characterised by a logarithmic normal function with a mass median aerodynamic diameter of 0.7 μm and GSD of 1.9. These particles were thought to be more favourable in terms of deep lung penetration than micronised drug particles. Hickey (1992) obtained elongated disodium cromoglycate particles after treatment with a hydrophobic molecule, lauric acid and the aerodynamic properties of the elongated particles were investigated using an inertial impactor. The equivalent diameters of the particles (i.e. the diameters of spherical particles that have the same volume as the elongated particles) depositing on each stage of the impactor were markedly higher than the use of nominal cut-off diameters of the stages. As the cut-off diameters were obtained by calibration with spherical particles, these results suggested that elongated particles with larger volumes or masses can deposit at the same stage as spherical particles having smaller volumes or masses. Each elongated particle can therefore be expected to carry more drug to the stage on which it deposits than an equivalent sphere. However, this phenomenon needs to be confirmed *in vivo* and, most importantly, in clinical practice.

3.6.5 Hygroscopic growth of particles in the airways

As discussed above, the size of inhaled particles is a dominant factor influencing their deposition profiles. Many drugs administered by inhalation are water-soluble, hygroscopic materials and will sorb the ubiquitous water vapour present within the warm and humid environment of the respiratory tract. Consequently, the sizes and densities of the particles will change after inhalation and this will eventually affect the sites of deposition, which are apparently different for those of the non-hygroscopic particles using the same inhalation manoeuvres.

The hygroscopic growth of atmospheric pollutants upon entry into the lung has long been a subject of interest and similar effects for therapeutic aerosols have all been well documented, especially those given by MDIs and nebulisers. From a physico-chemical point of view, when the pressure of the water vapour produced by a particle is less than that of the ambient atmosphere, water molecules will condense until the water pressure is the same as the ambient water vapour. For insoluble particles, the sorbed water molecules will remain on the particle surface to form a thin film that can rapidly reach the equilibrium pressure. Thus, the effect of humidity on the particle size of these materials may be negligible for most practical purposes. However, if the particle is water-soluble, a

CHAPTER 3

solution will form on its surface, the vapour pressure of which will always be less than that of a pure solvent. Water molecules will continue to condense until an equilibrium between the vapour pressures is reached. However, theoretically, it will never be obtained at a constant RH of 100% due to the difference in vapour pressure of a solution and pure water. Furthermore, the vapour pressure of a droplet is higher than that calculated on a planar surface, due to the higher specific surface area of the particles. Thus, a particle will always achieve an equilibrium diameter, d_e, which can be given by the following equation:

$$\frac{d_e}{d_o} = \sqrt[3]{\frac{\rho_o}{\rho_e} \left[1 + \frac{M_W iH}{M_o(K - H)} \right]} \qquad (3.12)$$

where d_o is the initial diameter of the particle, ρ_o is the initial density of the particle, ρ_e is the density at equilibrium, and M_W and M_o are the molecular weights of water and the particle, respectively (Ferron, 1977); H is the relative humidity, i the number of ions into which the material dissociates in water, and K a constant determined by the Kelvin equation:

$$K = \exp\left(\frac{4\sigma M_W}{RT\rho_e d_e}\right) \qquad (3.13)$$

where σ is the surface tension, d_e is the diameter at the equilibrium, R is the absolute gas constant and T is the absolute temperature. Thus, the final diameter of a water soluble particle is dependent upon both the initial diameter and the properties of the particulate materials as well as the temperature and humidity of the environment. The hygroscopic growth of particles can be assessed by growth ratio and rate. The former is the parameter used to assess the final diameter of a particle after the growth has reached equilibrium, i.e. d_e/d_o, whilst the latter is a dynamic parameter used to assess the rate at which the particle grows in size.

According to Xu and Yu (1985), the growth ratio of a particle is mainly determined by the initial particle diameter and the relative humidity of the environment. As is shown in Figure 3.6, the growth ratios of ultrafine particles (<0.1 μm) increase with initial particle diameters whilst the growth of particles greater than approximately 0.5 μm is essentially a constant, independent of the initial particle size. Furthermore, the growth ratio is drastically increased for small increases in relative humidity. However, other factors such as the hygroscopic nature may also be taken into account when considering the growth ratio of particles. Many studies have been published which deal with the growth ratios of inhalation aerosols, including theoretical calculation and experimental determination. For example, the growth ratio of disodium cromoglycate was calculated as 2.60, assuming a relative humidity of 99.5% and growth under equilibrium conditions (Gonda et al., 1981). However, the growth ratio of the same drug was measured experimentally to be 1.31, and smaller measured values have

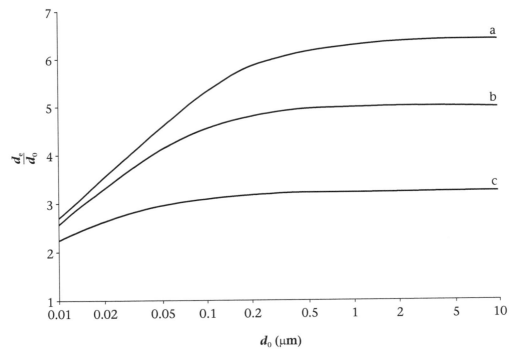

Figure 3.6: Effects of initial particle size and environmental relative humidities on the hygroscopic growth of particles (adapted from Xu and Yu, 1985).

been shown to be the case for other drugs. This may be partly due to the relative humidities at which the data were generated which were in the order of 90–98%, whilst the theoretical calculations used 99.5%.

As mentioned above, the hygroscopic growth of a particle is a dynamic process and its rate is also an important factor. It is dependent upon the rate of transport of water molecules to the surface and the removal of the heat associated with the condensation process. An empirical equation has been proposed to assess the growth rate of particles (Xu and Yu, 1985):

$$d_t - d_0/d_e - d_0 = 1 - \exp(-2.3t^{0.62}/d_0^{1.12}) \tag{3.14}$$

where d_t is the diameter of the particle at time t, after entering the humid atmosphere. Thus, the growth rate of a particle is largely dependent upon its initial diameter d_0, and the smaller the particles, the more quickly they will reach their equilibrium diameters. Morrow (1986) has calculated the growth rates of particles with diameters of 0.2, 0.6, 2.0, 6.0 and 10.0 μm in different generations of airways at inhalation rates of 6 and 60 l min^{-1}. Fine particles (<0.6 μm) were shown to attain their equilibrium diameters within a residence time of approximately 1 s but large particles could not reach equilibrium even after 10 s. Since a lower inhalation flow rate can increase the residence time of particles in each generation of respiratory tract, the growth of ultrafine particles may reach equilibrium before the tenth generation at $6 \, l \, min^{-1}$ but would not do so before the twentieth generation at $60 \, l \, min^{-1}$.

Since particle growth is determined by both the temperature and the relative humidity of the surrounding atmosphere in the respiratory tract, an accurate measurement of both factors is an essential prerequisite for determining the growth of hygroscopic particles. Although there have been many attempts to measure the temperature and humidity profiles in the respiratory tract, most of these methods have been limited as they are invasive and change the normal conditions. Further, the exact measurement sites within the lung during the test are often ambiguous. The temperature and humidity profiles in the lung were estimated for a range of physiologically-realistic breathing conditions and put into a format suitable for particle monitoring purposes. For oral breathing, the relative humidity (RH) in the trachea has been estimated to be 90% and RH values increase monotonically by 1% increments for each downstream bronchial passage until a saturation level of 99.5% is achieved after the tenth generation (the peripheral bronchioles). The temperature in the respiratory tract for most breathing conditions may be considered to be 37°C.

Studies on the hygroscopic growth of aerosols have been mainly related to the drugs delivered by MDIs and nebulisers. However, there have been several reports dealing with the hygroscopic growth of dry powders in the airways and two saturated fatty acids, lauric and capric acid, were employed to reduce the hygroscopic growth of aerosol dry powders. For example, disodium fluorescein (DF) dry powders were coated with lauric or capric acid by an adsorption/coacervation technique (Hickey *et al.*, 1988). The *in vitro* hygroscopic growth of the coated and uncoated powders were investigated under controlled temperature (37°C) and relative humidity (20 and 97% RH) (Hickey *et al.*, 1988, 1990). Particle size was characterised by a cascade impactor using a preconditioned air stream at a flow rate of $12.45 \, l \, min^{-1}$. The transit time of the aerosol at these controlled conditions was approximately 40 s. Disodium fluorescein particles exhibited a hygroscopic growth ratio of 1.5 at a relative humidity of 97%. This was reduced to 1.3 by coating with 0.15 g of lauric acid or 0.8 g of capric acid per gram of DF, and was eliminated with 0.2 g of lauric acid or 0.08 g of capric acid per gram of DF. The reduced growth ratio was attributed to the formation of a hydrophobic coat around the fluorescein particles and the increased capacity of such particles to adsorb water in order to form droplets whose vapour pressure is the same as that of the surrounding atmosphere. Predictably, the dissolution rate of DF after coating with the hydrophobic materials was found to decrease with increased surface coverage.

Particles of a hygroscopic drug, disodium cromoglycate (DSCG), was subjected to similar treatment. Thus, DSCG particles were coated with lauric acid by adsorption–coacervation from a methylene chloride/chloroform solution (Hickey,

1992). In this case, the possible change in the hygroscopic growth of the drug was found to be complicated by a concomitant change in the shape of coated particles. Elongated particles were produced after the coating process and, as discussed above, the geometric change may bring about some advantages for controlled lung delivery over the more spherical form, even in the absence of any hygroscopic effects.

The use of hydrophobic materials in the preparation of aerosol particles could have other beneficial effects. The coating of hygroscopic drugs with such materials will undoubtedly reduce the sorption of water by hygroscopic particles during preparation and storage. A reduction in the water uptake may offer many advantages such as improving the physico-chemical stability and preventing microbial growth. As mentioned previously, water uptake will increase the cohesion and adhesion of powders by generating increased capillary forces, decreasing the deaggregation and dispersion of drug particles in the air stream. Through a careful choice of coating materials, powders exhibiting the desired interparticulate forces and maximum physico-chemical stability, may be prepared to achieve optimum delivery of drugs to the lung.

In conclusion, many factors may play a role in determining the delivery efficiency of drugs by dry powder inhalers and the subsequent deposition profiles of the inhaled drug particles in the respiratory tract. These include the design of the inhaler devices, inhalation manoeuvre and powder formulation. The inhaler device should not only provide sufficient air turbulence to disperse and deaggregate drug particles but also ensure adequate flow rates of the inhaled air stream. The inspiratory efforts of the patient impart the energy to detach the drug from any carrier particles and break up any drug aggregates such that discrete drug particles that reach the lower airways will be generated. The inhalation mode also affects the overall deposition profiles of inhaled drug particles in the respiratory tract by means of altering some of the aerodynamic parameters of the airborne particles, particularly the particle momentum. Finally, the physico-chemical properties of the drug particles, including particle size, shape, density and hygroscopicity, all determine particle deposition in the airways.

REFERENCES

ADJEI, A.L. and GUPTA, P.K. (1997) Dry powder inhalation aerosols. In: ADJEI, A.L. and GUPTA, P.K. (eds), *Inhalation Delivery of Therapeutic Peptides and Proteins*. New York, Marcel Dekker, 625–665.

AGNEW, J.W. (1984) Physical properties and mechanisms of deposition of aerosols. In: CLARKE, S.W. and PAVIA, D. (eds), *Aerosols and the Lung: Clinical and Experimental Aspects*. London, Butterworth.

ALLEN, M.D. and RAABE, O.G. (1985) Slip correction measurements of spherical solid aerosol particles in an improved Millikan apparatus. *Aerosol Sci. Tech.* **4**, 269–286.

■ CHAPTER 3 ■

ANDERSON, M., PHILIPSON, K., SVARTENGREN, M. and CAMNER, P. (1995) Human deposition and clearance of 6-micron particles inhaled with an extremely low flow rate. *Exp. Lung Res.* **21**, 187–195.

AUTY, R.M., BROWN, K., NEALE, M.G. and SNASHALL, P.D. (1987) Respiratory tract deposition of sodium cromoglycate is highly dependent upon technique of inhalation using the Spinhaler. *Brit. J. Dis. Chest* **81**, 371–380.

BARON, P.A. and WILLEKE, K. (1992) Gas and particle motion. In: WILLEKE, K. and BARON, P.A. (eds), *Aerosol Measurement: Principles, Techniques, and Applications.* New York, Van Nostrand Reinhold, 23–40.

BELL, J.H. (1992) Dry powder inhalers. Innovation, performance assessment and the realities. *Management Forum*, London, 15 December.

BELL, J.H., HARTLEY, P.S. and COX, J.S.G. (1971) Dry powder aerosols I: A new powder inhalation device. *J. Pharm. Sci.* **60**, 1559–1564.

BOGAARD, J.M., SLINGERLAND, R. and VERBRAAK, A.F.M. (1989) Dose-effect relationship of terbutaline using a multi-dose powder inhalation system (Turbohaler) and salbutamol administered by powder inhalation (Rotahaler) in asthmatics. *Pharmatherapeutica* **5**, 400–406.

BRAIN, J.D., VALBERG, P.A. and SNEDDON, S. (1985) Mechanisms of aerosol deposition and clearance. In: MOREN, F., NEWHOUSE, M.T. and DOLOVICH, M.B. (eds), *Aerosols in Medicine: Principles, Diagnosis and Therapy.* Amsterdam, Elsevier Science Publishers, 123–148.

BROWN, P.H., NING, A.C.W.S., GREENING, A.P., MCLEAN, A. and CROMPTON, G.K. (1995) Peak inspiratory flow through Turbohaler® in acute asthma. *Eur. Respir. J.* **8**, 1940–1941.

BUTTERWORTH, S. (1986) The structure and function of inhalers. *Pharmacy Update*, November/December, 405–409.

BYRON, P.R. and PATTON, J.S. (1994) Drug delivery via the respiratory tract. *J. Aerosol Med.* **7**, 49–75.

CHAN, H.-K. and GONDA, I. (1989) Respirable form of crystals of cromoglycic acid. *J. Pharm. Sci.* **78**, 176–180.

CHAN, T.L. and YU, C.P. (1982) Charge effects on particle deposition in the human tracheobronchial tree. In: WALTON, W.H. (ed.), *Inhaled Particles.* V. Oxford, Pergamon.

CHEN, R.Y. (1978) Deposition of charged particles in tubes. *J. Aerosol Sci.* **9**, 449–463.

CLARK, A.R. and HOLLINGWORTH, A.M. (1993) Simulated 'in-use' powder inhaler resistance and peak inspiratory conditions in healthy volunteers – implications for in vitro testing. *J. Aerosol Medicine – Deposition, Clearance and Effects in the Lung* **6**, 99–110.

CLAY, M.M. and CLARKE, S.W. (1987) Wastage of drug from nebulizers: a review. *J. Roy. Soc. Med.* **80**, 38–39.

COLE, R.B. and MACKAY, A.D. (1990) *Essentials of Respiratory Disease.* Edinburgh, Churchill Livingstone.

CURRY, S.H., TAYLOR, A.J. and EVANS, S. (1975) Deposition of disodium cromoglycate administered in three particle sizes. *Brit. J. Clin. Pharmacol.* **2**, 257–270.

CYR, T.D., GRAHAM, S.J., LI, K.Y.R. and LOVERING E.G. (1991) Low first spray drug content in albuterol metered dose inhalers. *Pharmaceut. Res.* **8**, 658–660.

DAVIES, C.N. (1961) A formalized anatomy of the human respiratory tract. In: DAVIES C.N. (ed.), *Inhaled Particles and Vapours.* Oxford, Pergamon Press, 82–88.

DAVIES, P.J., HANLON, G.W. and MOLYNEUX, A.J. (1976) An investigation into the deposition of inhalation aerosol particles as a function of air flow rate in a modified 'Kirk lung'. *J. Pharm. Pharmacol.* **28**, 908–911.

DEBOER, A.H., WINTER, H.M.I. and LERK, C.F. (1996) Inhalation characteristics and their effects on in vitro drug delivery from dry powder inhalers. I. Inhalation characteristics, work of breathing and volunteers' preference in dependence of inhaler resistance. *Int. J. Pharm.* **130**, 231–244.

DOLOVICH, M., VANZIELEGHEM, M. and HIDINGER, K.G. (1988) Influence of inspiratory flow rate on the response of terbutaline sulphate inhaled via a Turbohaler®. *Am. Rev. Respir. Dis.* **137**, 433.

EDWARDS, D.A., HANES, J., CAPONETTI, G., HRKACH, J., BEN-JEBRIA, A., ESKEW, M.L., MINTZES, J., DEAVER, D., LOTAN, N. and LANGER, R. (1997) Large porous particles for pulmonary drug delivery. *Science* **66**, 1254–1258.

FERRON, G.A. (1977) The size of soluble aerosol particles as a function of the humidity of the air. Application to the human respiratory tract. *J. Aerosol Sci.* **8**, 251–267.

FULLER, R. (1995) Powder inhalers, past, present and future. *Proceedings of Drug Delivery to the Lung VI.*

GONDA, I. (1981) Study of the effects of polydispersity of aerosols on regional deposition in the respiratory tract. *J. Pharm. Pharmacol.* **33**, (Suppl.) 52P.

GONDA, I. (1992) Targeting by deposition. In: HICKEY, A.J. (ed.), *Pharmaceutical Inhalation Aerosol Technology.* New York, Dekker, 61–82.

CHAPTER 3

GONDA, I., KAYES, J.B., GROOM, C.V. and FILDES, F.J.T. (1981) Characterization of hygroscopic inhalation aerosols. In: STANLEY-WOOD, N. and ALLEN, T. (eds), *Particle Size Analysis*. New York, J. Wiley and Sons, 31–43.

GROSSMAN, J. (1994) The evolution of inhaler technology. *J. Asthma* **31**, 55–64.

HATCH, T.F. and GROSS, P. (1964) *Pulmonary Deposition and Retention of Inhaled Aerosols*. New York, Academic Press.

HEYDER, J., GEBHART, J., RUDOLPH, G., SCHILLER, C.F. and STAHLHOFEN, W. (1986) Deposition of particles in the human respiratory tract in the size range 0.005–15 μm. *J. Aerosol Sci.* **17**, 811–825.

HICKEY, A.J. (1992) Summary of common approaches to pharmaceutical aerosol administration. In: HICKEY, A.J. (ed.), *Pharmaceutical Inhalation Aerosol Technology*. New York, Marcel Dekker, 255–259.

HICKEY, A.J., GONDA, I., IRWIN, W.J. and FILDES, F.J.T. (1990) Effect of hydrophobic coating on the behaviour of a hydroscopic aerosol powder in an environment of controlled temperature and relative humidity. *J. Pharm. Sci.* **79**, 1009–1014.

HICKEY, A.J., JACKSON, G.V. and FILDES, F.J.T. (1988) Preparation and characterization of disodium fluorescein powders in association with lauric and capric acids. *J. Pharm. Sci.* **77**, 804–809.

HINDS, W.C. (1982) *Aerosol Technology: Properties, Behaviour and Measurement of Airborne Particles*. New York, Wiley Interscience.

HINDS, W.C. (1982) In: *Aerosol Technology*. New York, John Wiley and Sons.

HORSLEY, M. (1988) Nebuliser therapy. *Pharmaceut. J.* **240**, 22–24.

JOHN, W. (1980) Particle charge effects. In: WILLEKE, K. (ed.), *Generation of Aerosols and Facilities for Exposure Experiments*. Ann Arbor Science Publishers Inc., 141–151.

JOHNSON, C.E. (1989) Principles of nebulizer-delivered drug therapy for asthma. *Am. J. Hosp. Pharm.* **46**, 1845–1855.

JOHNSON, D.L. and MARTONEN, T.B. (1994) Behavior of inhaled fibers: potential applications to medicinal aerosols. *Particul. Sci. Technol.* **12**, 161–173.

MARTONEN, T.B. (1993) Mathematical model for the selective deposition of inhaled pharmaceuticals. *J. Pharm. Sci.* **82**, 1191–1199.

MARTONEN, T.B. and KATZ, I.M. (1993) Deposition patterns of aerosolized drugs within human lungs: effects of ventilatory parameters. *Pharmaceut. Res.* **10**, 871–878.

MARTONEN, T.B. and YANG, Y. (1996) Deposition mechanics of pharmaceutical particles in human airways. In: HICKEY, A.J. (ed.), *Inhalation Aerosols: Physical and Biological Basis for Therapy*. New York, Marcel Dekker, 3–27.

MOREN, F., NEWHOUSE, M.T. and DOLOVICH, M.B. (1985) *Aerosol in Medicine: Principles, Diagnosis and Therapy*. Amsterdam, Elsevier.

MORROW, P.E. (1986) Factors determining hygroscopic aerosol deposition in airways. *Physiol. Rev.* **66**, 330–376.

NEUMANN, B.S. (1967) The flow properties of powders. In: BEAN, H.S., BECKETT, A.H. and CARLESS, J.E. (eds), *Advances in Pharmaceutical Sciences*, Vol. 2. London, Academic Press, 181–221.

NEWMAN, S.P. and CLARKE, S.W. (1983) Therapeutic aerosols. I. Physical and practical considerations. *Thorax* **38**, 881–886.

NEWMAN, S.P., HOLLINGWORTH, A. and CLARK, A.R. (1994) Effect of different modes of inhalation on drug delivery from a dry powder inhaler. *Int. J. Pharm.* **102**, 127–132.

NEWMAN, S.P., MOREN, F., TROFAST, E., TALAEE, N. and CLARKE, S.W. (1991) Terbutaline sulphate Turbohaler®: effect of inhaled flow rate on drug deposition and efficacy. *Int. J. Pharm.* **74**, 209–213.

NEWMAN, S.P., PAVIA, D. and CLARKE, S.W. (1981) How should a pressurised beta-adrenergic bronchodilator be inhaled? *Eur. J. Resp.* **62**, 3–21.

NEWMAN, S.P., PAVIA, D., GARLAND, N. and CLARKE, S.W. (1982) Effects of various inhalation modes on the deposition of radioactive pressurised aerosols. *Eur. J. Resp.* **63** (Suppl. 119), 57–65.

NEWMAN, S.P., PAVIA, D., MOREN, F., SHEAHAN, N.F. and CLARKE, S.W. (1981) Deposition of pressurized aerosols in the human respiratory tract. *Thorax* **36**, 52–55.

OLSSON, B. and ASKING, L. (1994) Critical aspects of the function of inspiratory flow driven inhalers. *J. Aerosol Med. – Deposition, Clearance and Effects in the Lung* **7** (Suppl. 1), S43–S47.

OWER, E. and PANKHURST, R.C. (1977) *The Measurement of Airflow*. Fifth edition. Oxford, Pergamon Press.

PEDERSEN, S. (1986) How to use a Rotahaler. *Arch. Dis. Child.* **61**, 11–14.

PEDERSEN, S., STEFFENSEN, G. (1986) Fenoterol powder inhaler technique in children – influence of inspiratory flow-rate and breath-holding. *Eur. J. Respir. Dis.* **68**, 207–214.

PEDLEY, T.J., SCHROTER, R.C. and SUDLOW, M.F. (1977) Gas flow and mixing in the airways.

CHAPTER 3

In: WEST, J.B. (ed.), *Bioengineering Aspects of the Lung*. New York, Marcel Dekker, 113–125.

PERSSON, G., GRUVSTAD, E. and STAHL, E. (1988) A new multiple dose powder inhaler (Turbohaler®), compared with a pressurized inhaler in a study of terbutaline in asthmatics. *Eur. Respir. J.* **1**, 681–684.

PERSSON, G., OLSSON, B. and SOLIMAN, S. (1997) The impact of inspiratory effort on inspiratory flow through Turbohaler® in asthmatic patients. *Eur. Resp. J.* **10**, 681–684.

PITCAIRN, G., LUNGHETTI, G., VENTURA, P. and NEWMAN, S. (1994) A comparison of the lung deposition of salbutamol inhaled from a new dry powder inhaler at two inhaled flow rates. *Int. J. Pharm.* **102**, 11–18.

RAABE, O.G. (1978) A general method for fitting size distributions to multicomponent aerosol data using weighted least-squares. *Env. Sci. Technol.* **12**, 1162–1167.

REES, P.J., CLARK, T.J.H. and MOREN, F. (1982) The importance of particle size in response to inhaled bronchodilators. *Eur. J. Resp.* **63** (Suppl. 119), 73–78.

RILEY, D.J., LIU, R.T. and EDELMAN, N.H. (1979) Enhanced responses to aerosolised bronchodilator therapy in asthma using respiratory manoeuvres. *Chest* **76**, 501–507.

RILEY, D.J., WEITZ, B.W. and EDELMAN, N.H. (1976) The response of asthmatic subjects to isoproterenol inhaled at differing lung volumes. *Am. R. Resp. D.* **114**, 509–515.

SCHULTZ, R.K., MILLER, N.C., SMITH, D.K. and ROSS, D.L. (1992) Powder aerosols with auxiliary means of dispersion. *J. Biopharm. Sci.* **3**, 115–121.

SETHI, D.K., NORWOOD, D.L., HAYWOOD, P.A. and PRIME, D. (1992) Impact of extractable testing on MDI development programs. *J. Biopharm. Sci.* **3**, 63–68.

SMITH, G., HILLER, C., MAZUMDER, M. and BONE, R. (1980) Aerodynamic size distribution of cromolyn sodium at ambient and airway humidity. *Am. R. Resp. D.* **121**, 513–517.

SPIRO, S.G., BIDDISCOMBE, M., MARRIOTT, R.J., SHORT, M. and TAYLOR, A.J. (1992) Inspiratory flow rates attained by asthmatic patients through a metered dose inhaler and a Diskhaler inhaler. *Brit. J. Clin. Res.* **3**, 115–116.

SUDLOW, M.F., OLSON, D.E. and SCHROTER, R.C. (1971) Fluid mechanics of bronchial airflow. In: WALTON, W.H. (ed.), *Inhaled Particles III, England*. Unwin Bros Ltd., Surrey, UK. Gresham Press, 19–31.

SUMBY, B.S., COOPER, S.M. and SMITH, I.J. (1992) A comparison of the inspiratory effort required to operate the Diskhaler inhaler and Turbohaler inhaler in the administration of powder drug formulations. *Brit. J. Clin. Res.* **3**, 117–123.

THE TASK GROUP ON LUNG DYNAMICS (1966) Deposition and retention models for internal dosimetry of the human respiratory tract. *Health Phys.* **12**, 173–209.

THEIL, C.G. (1996) From Susie's question to CFC free: an inventor's perspective on forty years of metered dose inhaler development and regulation. *Respiratory Drug Delivery V* 115–123.

TIMBRELL, V. (1965) The inhalation of fibrous dusts. *Ann. NY Acad. Sci.* **132**, 255–273.

TIMSINA, M.P., MARTIN, G.P., MARRIOTT, C., GANDERTON, D. and YIANNESKIS, M. (1994) Drug delivery to the respiratory tract using dry powder inhalers. *Int. J. Pharm.* **101**, 1–13.

TIMSINA, M.P., MARTIN, G.P., MARRIOTT, C., LEE, K.C., SUEN, K.O. and YIANNESKIS, M. (1993) Studies on the measurement of peak inspiratory flow rate (PIF) with dry powder inhaler devices in healthy volunteers. *Thorax* **48**, 433.

WALL, D.A. (1995) Pulmonary absorption of peptides and proteins. *Drug Del.* **2**, 1–20.

WEIBEL, E.R. (1963) *Morphometry of the Human Lung*. Berlin, Springer.

XU, G.B. and YU, C.P. (1985) Theoretical lung deposition of hygroscopic NaCl aerosol. *Aerosol Sci. Tech.* **4**, 455–461.

ZANEN, P., VAN SPIEGEL, P.I., VAN DER KOLK, H., TUSHUIZEN, E. and ENTHOVEN, R. (1992) The effect of the inhalation flow on the performance of a dry powder inhalation system. *Int. J. Pharm.* **81**, 199–203.

■ CHAPTER 3 ■

Particle Interactions with Air Streams

Contents

4.1 INTRODUCTION

The interaction between particles and an air stream is important in many industrial processes such as pneumatic transport, particle fluidisation and dispersion. Fluidisation refers to a laterally confined mass of particles that are agitated by a rising stream of gas so that the particles are partly airborne and partly undergoing momentary contact with one another. Under certain conditions the gas-particle mass behaves effectively as a fluid and hence this phenomenon is called fluidisation. When particles are individually induced to move along with a stream of air passing through a conduit, they are said to be pneumatically transported. Under such conditions, the particles are relatively farther apart from one another than in the case of fluidisation. Dispersion of particles in an air stream usually refers to particle entrainment as discrete entities into air and can be regarded as fluidisation at the highest level of efficiency. It is important to appreciate that it is often very difficult to distinguish such processes from one another. If a powder is introduced into an air stream, some component particles may undergo fluidisation, others pneumatic transport and the rest dispersion. Whether a bulk powder is fluidised or dispersed is largely dependent upon the particle and fluid properties, and interaction between the particles and the fluid. For example, when a vertical flow generates a pressure gradient that exceeds the weight of a powder bed, then fluidisation of the powder is likely to occur. Increasing the fluid velocity will produce a higher pressure gradient and hence the flow may carry particles or their aggregates toward the flow direction of the air stream. If the flow velocity is further increased such that sufficient energy is generated to deaggregate airborne particles, discrete particles may be dispersed into the air stream with minimum interactions between each other. However, powder fluidisation may involve either pneumatic transport or dispersion of a portion of the powder bed whilst powder dispersion is almost always developed as a result of fluidisation.

Fluidisation, pneumatic transport and dispersion are important in processing many pharmaceutical powders. For example, fluidisation has provided one of the most efficient mechanisms for the drying of granular solids since it is more efficient and faster than that which can be achieved by conventional drying such as in a tray dryer (Rankell *et al.*, 1986). During fluidisation, each particle is completely surrounded by the drying gas, leading to a more efficient heat exchange between the particles and the surrounding environments. Fluidisation also produces intense mixing of component particles within the powder, resulting in uniform conditions of both temperature and particle size distribution throughout the bed.

The mechanisms of pneumatic transport have also been employed extensively in the pharmaceutical industry, one of the most familiar techniques being spray-drying (Rankell *et al.*, 1986). This involves atomising liquid droplets in a hot air stream, which carries them to a cyclone separation chamber. During pneumatic transport, rapid evaporation of the solvent results in the drying of the liquid droplets. Spray-drying is similar to fluidised bed drying in that both employ moving hot air to facilitate the drying process. However, spray-drying is different from fluidised bed drying in that the former involves the atomisation of a liquid into droplets whilst the latter directly fluidises solid particles. Particle dispersion in a liquid, a process used to prepare a pharmaceutical suspension, has been dealt with most extensively in many textbooks (e.g. Patel *et al.*, 1986; Buckton, 1995). Dispersion of particles in an air stream has also been an active field of research in powder technology (Kousaka, 1991). However, the implications of such research have not been fully appreciated by formulation scientists. Particle dispersion in air streams is an important phenomenon in powder formulation especially when developing medicinal powder aerosols.

As mentioned before, drug particles for inhalation are usually required to be within the particle size range 1.0–5.0 μm in diameter. These fine particles exhibit interparticulate interactions which are much more pronounced than the gravity force due to their low mass. The strong cohesive forces may be extremely difficult to overcome, leading to an increased aggregation and poor flow and dispersion properties. In order to solve these problems, drug particles are often mixed with coarser carrier particles, such as lactose (Ganderton, 1992). The carrier particles are designed to have such a size (63–90 μm being typical) that after inhalation, they will be retained in the inhaler device or deposit in the upper airways, mostly in the mouth and on the back of the throat. In order for the drug particles to reach the lower airways, they must therefore dissociate from the carriers and become dispersed in the inhaled air stream (Figure 4.1).

In reality, it is however very difficult to obtain an ordered mixture where discrete drug particles adhere to coarse carrier particles and a significant portion of drug particles may adhere to the carrier surface as aggregates (Ganderton and Kassem, 1992). Such aggregates of drug, whether adhered to a carrier surface or not, will require further dispersion to produce individual drug particles that are of a suitable size for deep lung penetration. Both drug aggregation and adhesion to the carrier particles are determined by interparticulate forces. Higher interparticulate forces not only produce more drug aggregates but also result in poorer drug deaggregation and dispersion. Furthermore, the properties of the air stream are also important in determining the deaggregation and dispersion of drug particles. A complete understanding of the various factors that participate in particle entrainment and dispersion is essential in the formulation of efficient medicinal aerosols. Therefore, it is the aim of this chapter to review some of the basic theories related to the dispersion of particles in a gaseous flow, derived by other scientific disciplines such as chemical engineering. Also discussed are the factors affecting re-entrainment and dispersion properties of particulate materials, especially those having a size range relevant to medicinal aerosols.

(A) Fine drug particles and coarse carrier particle before mixing

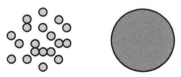

(B) The adhesion of fine drug particles to coarse carrier particles after mixing

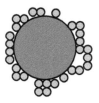

(C) Dissociation of fine drug particles from carrier particle after inhalation

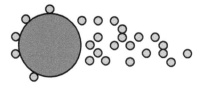

Figure 4.1: Schematic representation of adhesion and dissociation of fine particles.

4.2 DRAG FORCES OF AIR STREAMS

If a particle is placed in an air stream, or an air stream is blown over a particle, then it will be constantly bombarded from the direction of the gas flow by a large number of gas molecules, producing a drag force on the particle. The drag force created by air flow around a particle is a function of both the flow and the particle properties. For large particles, the gaseous flow may then be considered to be continuous whilst for very small particles, gaseous flow may no longer be a continuum. If a particle has a size less than the mean free path, λ, which is the mean distance a gas molecule travels before colliding with another gas molecule, then the motion of the particle is no longer determined by continuum flow considerations. Whether a particle is subject to a continuous or discontinuous flow depends upon the Knudsen number, K_n, which relates the gas molecular mean free path to the physical dimension of the particle:

$$K_n = \frac{2\lambda}{d_p} \tag{4.1}$$

where d_p is the diameter of the particle in question and λ is the mean free path of a molecule in air, which, at normal temperature and pressure is 0.0665 μm (Rader, 1990).

$K_n \ll 1$ indicates continuum flow whilst $K_n \gg 1$ indicates free molecular flow. The intermediate range, around 0.4–20, is usually referred to as the transition or slip flow regime.

If a spherical particle is suspended in a steady air stream that can be considered to be continuous with the particle, the aerodynamic drag force (F_{drag}) on the particle is proportional to both the cross-sectional area of the particle ($\pi d_p^2/4$) and the dynamic pressure acting on the particle from the air stream ($\rho_g V^2/2$). It can then be calculated as (Baron and Willeke, 1993):

$$F_{drag} = C_d \frac{\pi d_p^2}{4} \frac{\rho_g V^2}{2} = \frac{\pi}{8} C_d \rho_g V^2 d_p^2 \qquad (4.2)$$

where d_p is the particle diameter; V is the velocity of the free stream, ρ_g is the gas density and C_d is the drag coefficient.

Drug particles for inhalation usually have a particle size between 2 and 5 μm, which corresponds to values of the Knudsen number between 0.0665 and 0.0266 under normal conditions. Therefore, the air stream can be regarded as being continuous for such particles. However, most inhaled drug particles are irregularly-shaped and hence, equation (4.2) must be modified to give:

$$F_{drag} = \frac{\pi}{8} C_d \chi \rho_g V^2 d_p^2 \qquad (4.3)$$

where χ is the dynamic shape factor of the particle. The latter usually has a value >1 for non-spherical particles and more anisometric ones usually have a higher shape factor than those that are more isometric.

If a particle is so small that its Knudsen number is more than 1, equation (4.3) will have to be modified. In this case, the particle may travel a very small distance without collisions with any gas molecules and the particle is said to 'slip' between the gas molecules. Therefore, drag forces on the particles should be smaller than the drag force predicted by equation (4.3). To accommodate the difference, a slip correction factor, C_c, also referred to as the 'Cunningham slip correction factor', is introduced into the equation which becomes:

$$F_{drag} = \frac{\pi}{8C_c} C_d \chi \rho_g V^2 d_p^2 \qquad (4.4)$$

The Cunningham slip correction factor, C_c takes unity value for large particles but is >1 for small particles. An empirical fit for C_c has been reported (Allen and Raabe, 1985):

$$C_c = 1 + K_n \left(\alpha + \beta \exp \left(- \frac{\gamma}{K_n} \right) \right) \qquad (4.5)$$

where α, β and γ are constants for a specific particle.

■ CHAPTER 4 ■

The drag coefficient, C_d is a function of the Reynolds number of the particle, R_{ep}:

$$R_{ep} = \frac{d_p \rho_g V}{\eta} \tag{4.6}$$

where d_p, ρ_g and V are defined as above and η is the fluid viscosity.

Figure 4.2 shows the relationship between drag coefficient and the Reynolds number of the particle. If the Reynolds number of the particle is less than 0.1, then the particle is said to be in laminar flow. When the Reynolds number is over 1000, then the particle is said to be in a turbulent flow. When the Reynolds number falls between 0.1 and 1000, the particle is considered to be in the transitional region. If particles are exposed to different conditions as represented by the regions, then the drag coefficient will obviously vary.

In the laminar region where $R_{ep} < 0.1$, drag coefficient is then expressed as:

$$C_d = \frac{24}{R_{ep}} \tag{4.7}$$

Combining equations (4.4), (4.6) and (4.7) gives:

$$F_{drag} = \frac{3\pi\chi\eta V d_p}{C_c} \tag{4.8}$$

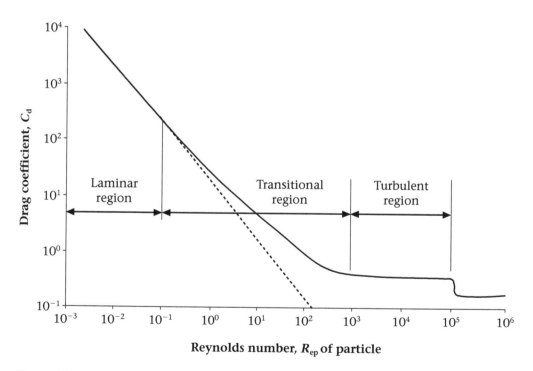

Figure 4.2: Drag coefficient as a function of the Reynolds number for spherical particles.

This equation is also known as Stokes law (see Section 3.2.2). In order to obey Stokes law laminar flow should prevail and the drag on the particle will then depend only on the particle-gas viscosity, η, gas velocity, V, particle diameter d_p and particle shape factor, χ.

Equation (4.7) assumes that the flow around the particle is laminar. As particles move faster (i.e. have a R_{ep} larger than 0.1), empirical modifications have to be made to the Stokes law (equations (4.9) and (4.10)).

If $0.1 \leq R_{ep} < 5$, then C_d is calculated as (Sartor and Abbott, 1973):

$$C_d = \frac{24}{R_{ep}} (1 + 0.0916R_{ep}) \tag{4.9}$$

If $5 \leq R_{ep} < 1000$, then the following empirical equation can be employed to calculate the C_d value (Friedlander, 1977):

$$C_d = \frac{24}{R_{ep}} (1 + 0.158R_{ep}^{2/3}) \tag{4.10}$$

When $1000 < R_{ep} < 10\,000$, then the drag coefficient becomes more or less a constant, approximately 0.42 (Heywood, 1962).

It has to be noted that all the above relationships between drag coefficient and particle Reynolds number were generated for smooth, spherical particles, and those with other morphologies exhibit different relationships between C_d and R_{ep} (Haider and Levenspeil, 1989). Particles with extreme shapes may have a significantly different drag coefficient from spherical particles. For example, fibres may have a drag coefficient which is up to four times lower than that of spheres when R_{ep} is of the order of 100 (Clift *et al.*, 1978).

Thus, the drag force of an air stream on a particle is a function of both particle size, shape and flow properties. Increasing the viscosity of the medium will increase the drag force of the stream on particles. This is very easy to accept as water exerts a much higher drag force on solid objects than an air stream. Increasing the velocity can also increase drag forces, which is equally apparent. It should be noted, however, that the relevant particle diameter in equations (4.3), (4.4) and (4.8) is that of the cross-sectional area of the particle vertical to the stream direction, which may not represent the true diameter of the particles.

4.3 PARTICLE ENTRAINMENT

During inhalation, dry powders may undergo two major processes, namely, entrainment of the ordered mixes of drug and carrier particles in the air stream and detachment of drug from the airborne carrier. The first step involves the fluidisation of dry powders from the surface of the inhaler chamber where the powders were originally packaged or released. This process resembles particle detachment from a stationary surface and will be discussed according to the basic theory generated from related fields. After dispersion, drug particles will be detached from the carrier, due to the turbulence of the air stream and the different aerodynamic properties of the large carrier and fine drug.

The entrainment of dry powders for inhalation is mainly dependent upon the drag forces acting on the particles from the inhaled air stream. If the drag forces can overcome particle–particle and particle–surface interactions as well as the gravitational forces of the particles, the latter will be detached and entrained into the air stream. The drag forces of an air stream on an individual particle are very difficult, if not impossible, to measure accurately. However, insight into fluid–particle interactions in particle entrainment may be gained from the treatment of a particle suspended in a steady air stream.

4.3.1   Particle detachment from stationary surfaces

The detachment of particles from their adhered surfaces usually occurs by three mechanisms, rolling, sliding and lifting (Soltani *et al.*, 1995).

A particle will be removed or dislodged from its adhered surface if the drag force overcomes the adhesion forces and the weight of the particle (Figure 4.3).

A particle will start to roll along the surface, if:

$$F_{\mathrm{drag}} \geq (F_{\mathrm{ad}} + W) \, \frac{r_{\mathrm{c}}}{r - r_{\mathrm{c}}} \qquad (4.11)$$

where F_{drag} is the drag force, F_{ad} is the overall adhesion force, W is the particle weight, and r and r_{c} are the radii of the particle and contact area respectively.

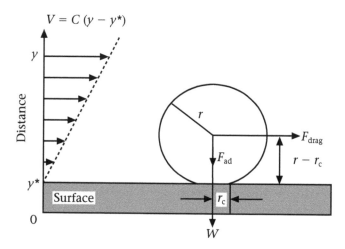

Figure 4.3: Relationship between adhesion force, drag force and geometry of the particle.

A particle will slide or be lifted, if the following conditions occur:

$$F_{drag} \geq \mu(F_{ad} + W) \tag{4.12}$$

where μ is the coefficient of friction.

For fine particles, when $F_{ad} \gg W$, then:

$$F_{drag} \geq \mu F_{ad} \tag{4.13}$$

Although the drag force acting on a particle in laminar flow can be calculated from equation (4.8), the gas velocity, included as a term in this equation, is the velocity at its centre. The flow will vary in velocity at different levels of the particle, relative to the y axis in Figure 4.3. For a linear velocity distribution in a laminar layer, the flow velocity will increase linearly from 0, at the contact point between the particle and surface where $y = y^*$, to $C(d - r_c - y^*)$ at the top of the particle, where d is the particle diameter, y^* is the thickness of the surface and C is a constant related to the flow. Therefore, the drag forces exerted on a particle will vary with the velocity of the air stream acting on the relative level of the particle. All these forces have to be summed to achieve the total drag force of the air stream acting on a particle. A more complicated mathematical treatment needs to be adopted involving integrating drag forces from the contact point through the particle centre to the top of the particle, but this is beyond the scope of this work. Interested readers should refer to relevant articles (e.g. Zimon, 1982; Matsusaka and Masuda, 1996).

4.3.2 Powder fluidisation

Powder fluidisation may take place when an air stream is introduced uniformly across the base of a bed of powder and is allowed to flow upward through the bed (Figure 4.4). Before the fluidisation conditions are achieved, the powder bed may expand upwards due to the drag forces of the air stream. Increasing the air flow velocity will produce a sufficient pressure gradient across the powder bed to overcome the weight and internal attractive and frictional forces. The powder bed will be expanded to such an extent that constituent particles will separate from each other progressively until the void spaces become so large that each individual particle or agglomerate begins to circulate locally. At this point, fluidisation is said to have started. A further increase in the gas velocity will produce an overall circulation of the powder bed and the powder will be carried upwards by the rising air stream.

There are three major criteria which can be used to describe the fluidisation process, namely, pressure gradient across the powder bed, expansion of the bed and fluid velocity. The pressure gradient is expressed as $\Delta p/Li$, where Δp is the pressure gradient and Li

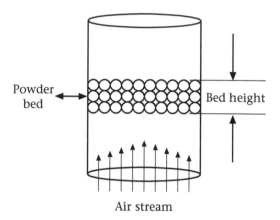

Figure 4.4: Fluidisation of a powder bed.

is the initial bed height. The bed expansion can be quantified by the void fraction ϵ, which is in turn calculated by $(V_b - V_t)/V_b$, where V_b is the bulk volume of the powder bed and V_t is the total volume occupied by each constituent particle. Although powders have completely different fluidisation properties, most fluidisation processes will exhibit typical interrelationships, as shown in Figure 4.5, the left hand plot representing the change in the pressure gradient with the flow velocity and the right hand plot showing the concomitant change in the bed void fraction (Hammond, 1958).

It can be seen from Figure 4.5 that by increasing the fluid velocity from A to B, the pressure gradient across the powder bed is increased almost linearly whilst the void fraction of the bed remains more or less the same. This is the period when the drag forces of the air flow are too low to exert any significant change in the packing of the powder bed. A further increase in flow velocity from B to C results in a deviation from linearity of the pressure gradient. Here, the air drag forces are sufficient to overcome interparticulate forces within the powder bed such that the constituent particles begin to rearrange themselves in order to provide the least possible area in the direction of the flow. Since significant bed expansion has not occurred, there is little change in the bed void fraction when the flow velocity is increased up to the point C. Therefore, the powder bed is said to be a 'fixed-bed' from A to C. Further increasing the fluid velocity from point C will terminate the 'fixed-bed' conditions.

Particles and/or aggregates will start to move upwards since the total drag forces of the flow will exceed the gravitational forces of the powder bed due to its mass. The bed begins to expand as fluidisation begins. Hence, the void fraction of the bed is increased, i.e. the bulk volume of the bed is increased. The particles begin to separate from one another, providing an abrupt increase in the void area in the flow direction, which results in a rapid drop in the pressure gradient from C to D. After this, pressure gradients generally remain constant, as indicated in the left of Figure 4.5, from D to E. However,

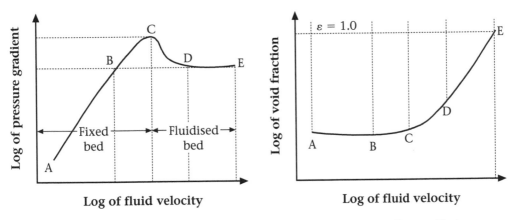

Figure 4.5: Typical relationships between fluid velocity and the pressure gradient and between fluid velocity and void fraction of powder bed.

bed expansion will continue until the void fraction of the bed increases to 1 (from D to E in the right hand curve of Figure 4.5), when all the particles are dispersed in the fluid. The difference in the pressure gradient from C to D is a characteristic of interparticulate forces in the powder bed. Higher interparticulate forces within the powder will require higher drag forces, i.e. a bigger pressure gradient across the powder bed, to initiate particle rearrangement within the powder and vice versa. In order to obtain satisfactory fluidisation, the smaller the difference in the pressure gradients from C to D the better. In summary, less cohesive powders exhibit better fluidisation properties than more cohesive powders.

Fluidisation can be either 'aggregative' or 'dispersive', depending upon the fluid and powder properties. In 'aggregative' fluidisation, clusters, aggregates or agglomerates of particles circulate in the bed and are carried upwards by the fluid whilst in 'dispersive' fluidisation, most of the particles are deaggregated or dispersed into single particles and the fluid phase is uniformly distributed. Therefore, the difference between aggregative and dispersive fluidisation is actually a matter of the degree of dispersion. In the majority of cases, dispersive fluidisation is preferred to aggregative fluidisation and this is often achieved when the fluid employed is a liquid. However, it can also be approached with a gas as the fluid, provided the particles are free flowing and sufficient drag forces are generated to overcome both the gravitational and interparticulate forces within the fluidised bed.

Particle size, size distribution, particle shape and density are among the most important factors in determining fluidisation properties. For example, free-flowing particles between 20 and 80 μm in diameter are generally most satisfactorily fluidised. Powders have been classified into groups based upon their particle size and density with a view to predicting their fluidisation properties at ambient conditions (Table 4.1).

TABLE 4.1

Powder classification (Geldart, 1972)

Groups	Particle size (µm)	Density (g cm^{-3})	General properties
A	20–100	<1.4	Aeratable, circulate when fluidised
B	40–500	1.4–5	Sand-like, readily fluidisable
C	<30		Cohesive and difficult to fluidise
D	>600		May be spouted[a]

[a]A spouted bed is formed within a cone-bottomed column having an opening at the apex of the cone for gas introduction. At a sufficient gas velocity, a stream of gas and solid particles rises in a central core, or spout, that bursts, fountain-like, through the upper surface of the bed.

Group A powders generally have intermediate particle size, low density, and often have a porous structure. Powders in this group are mildly cohesive and their interparticulate forces are of a similar order to the gravitational forces due to their mass. During fluidisation of group A powders, gross powder circulation occurs and the fluidised bed collapses slowly after the air supply is cut off. Group B powders are of a larger particle size and higher density. They are thus readily fluidisable but little powder circulation occurs during fluidisation or entrainment and the bed collapses rapidly after the external force is removed. Group C powders have a smaller particle size and their interparticulate forces are usually greater than the gravitational forces so such powders are highly cohesive and very difficult to fluidise or entrain in an air stream. This is because the interparticulate forces are greater than the gravitational forces and also because the drag forces which a gas stream might exert on the particles are not large enough to induce fluidisation. As soon as a critical velocity of the air stream has been reached, the particles often exhibit an erratic plug entrainment behaviour, where agglomerates of particles are entrained intermittently. In contrast, group D powders are coarse and easily flowable.

Medicinal aerosols usually have a diameter less than 7 µm and belong to group C. Consequently, they are sticky and difficult to fluidise when entrained in an air stream. This explains the need to employ carrier particles 63–90 µm diameter to aid the flowability of such drug powders. The carrier particles belong to group B powders and will thus improve both the fluidisation and entrainment of the fine drug powder.

4.4 PARTICLE DISPERSION

The dispersion of particles in an air stream is important in many pharmaceutical industrial processes such as pneumatic transport, size classification and reduction. The best dispersions, i.e. those with a large percentage of the particulates as discrete entities, are almost always obtained by passing a metered quantity of the particulate material into a

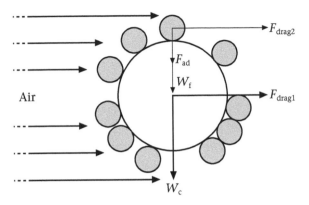

Figure 4.6: Diagram depicting the detachment of fine particles from an airborne carrier particle (see text for a key to the symbols).

gas stream through a nozzle of some description. Numerous reports have been published reporting the detachment of dusts and powders from stationary surfaces (Corn, 1966; Zimon, 1982) and the dispersion of aggregates of similar particles in an air stream (Endo *et al.*, 1997). However, very little work has been carried out to investigate the dispersion of aggregates composed of fine particles adhered to coarse carrier particles, such as that which exists in a typical dry powder aerosol formulation. Such a process can be expected to be much more complicated than either powder detachment from a stationary surface or dispersion of aggregates composed of primary particles of similar morphology since both the carrier and fine particles are exposed to a variety of different forces, some of which are shown in Figure 4.6.

If a particle is placed in a laminar flow stream and it is only subject to a horizontal air drag force, F_{drag}, and gravitational force due to its mass, W, then the acceleration of the particle (dV_p/dt) can be calculated as:

$$\frac{dV_p}{dt} = \frac{F_{drag}}{W} = \frac{3\pi\chi\eta V d_p}{\frac{\pi}{6}d_p^3\rho_p} = \frac{18\chi\eta V}{d_p^2\rho_p} \tag{4.14}$$

where V_p is the particle velocity, W is the mass of the particle, and the rest of the factors are defined above.

It can be seen from equation (4.14) that the acceleration of a particle decreases as particle diameter increases. Smaller particles will therefore achieve higher acceleration than larger particles under similar flow conditions. It is the difference in acceleration that provides the driving force for fine particles to detach from the coarse carrier particle. A fine particle will be detached from the coarse particle, if:

$$(F_{drag2} - \mu F_{ad})/W_f > F_{drag1}/W_c \tag{4.15}$$

■ CHAPTER 4 ■

or:

$$F_{drag2} > \mu F_{ad} \text{ when } W_f \ll W_c \tag{4.16}$$

where F_{drag1} and F_{drag2} are the drag forces for the coarse and fine particles, respectively, W_f and W_c are the weights of the fine and coarse particles, respectively and F_{ad} is the adhesion force between the fine and coarse particles. It is obvious that fine particles that are positioned at the top or the bottom of carrier particles are more likely to detach from the carrier than particles positioned at the back or front (Zimon, 1982). However, airborne particles are known to rotate constantly in an air stream and hence, most of the adhered particles may be, at some point in time, oriented in a position favourable for detachment. Further, this spin generates centrifugal forces on adhered particles, which also assists particle detachment, but in a complicated manner. Particle dispersion may also be promoted by the turbulence of the air stream. Local turbulence at the sites of particle adherence will be induced due to the change in the direction of the air stream in these areas. Such a change in direction leads to lifting forces being generated which, if they exceed the adhesion forces, will lead to the fine particles being detached from the carrier.

4.5 FACTORS AFFECTING ENTRAINMENT AND DISPERSION OF PARTICLES IN AIR STREAMS

As discussed above, many factors may affect entrainment and dispersion properties of particulate materials in an air stream, including air flow properties, particle properties, interaction between particles and air flow, and interparticulate forces within the powders. In terms of the air flow properties, the velocity is a dominant factor since it determines both the turbulence and drag forces acting on the particle. As far as particle properties are concerned, the particle size is of primary importance in determining both the aerodynamic properties, the interparticulate interactions, and the particle interactions with the air stream. The significance of interparticulate forces in particle dispersion also deserves much attention since they participate in nearly all the steps from fluidisation to dispersion.

4.5.1 Fluid properties

The most important single parameter used to characterise flow properties relevant to interaction with a particle is the Reynolds number, R_e

$$R_e = \frac{\rho_g V d'}{\eta} \tag{4.17}$$

where V is the velocity of the fluid, ρ_g is the gas density, η is the gas dynamic viscosity and d' is the characteristic dimension of the object, such as the diameter of a duct through which the fluid travels.

As has been mentioned earlier (Section 3.2.2) at low Reynolds numbers, the flow is streamline or laminar where frictional forces dominate the flow. Laminar flow occurs in a circular duct when the Reynolds number is less than about 2000. With increasing Reynolds number, i.e. at higher flow velocity, the flow becomes increasingly turbulent with the formation of many vortices and eddies. Generally, turbulent flow conditions are considered to have developed when the flow Reynolds number exceeds 4000. In the intermediate transitional range, the gas flow can be either laminar or turbulent, depending upon the previous history of the gas motion. For instance, if the gas velocity is increased into this intermediate region slowly, the flow may remain laminar. On the other hand, if the gas flow is decreased into this region, then the flow is most likely to be turbulent.

Increasing the flow velocity will not only increase the turbulence of an air stream, but also the drag forces of the fluid acting on a particle located in the direction of flow. Since these two factors are the major driving forces leading to re-entrainment and dispersion, the imposition of higher flow velocities almost always leads to both the removal of particles from stationary surfaces and their detachment from airborne carrier particles. This is of great importance with dry powder aerosols where the flow rate of the inhaled air stream generated by patients provides the primary energy for aerosolisation and deaggregation of drug particles.

Increasing the flow rate of a fluid stream to improve particle removal from a stationary surface is a well-known practice in powder technology. For example, the removal efficiency of latex (styrene/divinylbenzene) particles 1–10 μm in diameter from a glass surface, with a high-speed air jet, was found to be directly related to the dynamic pressure, which in turn was proportional to the square of the linear flow rate of the air stream (Gotoh *et al.*, 1994a). These workers found a critical dynamic pressure below which no significant particle removal was observed. Beyond this, a further increase in the dynamic pressure greatly increased removal efficiency. For example, the removal of adhered particles of 5.7 μm in diameter did not commence until the dynamic pressure reached about 4×10^4 Pa. A further slight increase in dynamic pressure of the air stream resulted in the removal efficiency increasing drastically until more than 90% of particles were removed at about 5×10^4 Pa (Gotoh *et al.*, 1994a).

Similarly, increasing the flow rates through an inhaler has also been found to improve particle detachment from device components, producing a higher emitted dose (Hindle *et al.*, 1994). For example, the doses emitted from most marketed dry powder inhalers were found to be higher after inhalation at $100\,l\,min^{-1}$ for 2.4 s than after inhalation at $60\,l\,min^{-1}$ for 4 s despite the inhalation volume being maintained at 4 l (Figure 4.7).

The flow rate is also critical in the deaggregation of entrained particle agglomerates.

CHAPTER 4

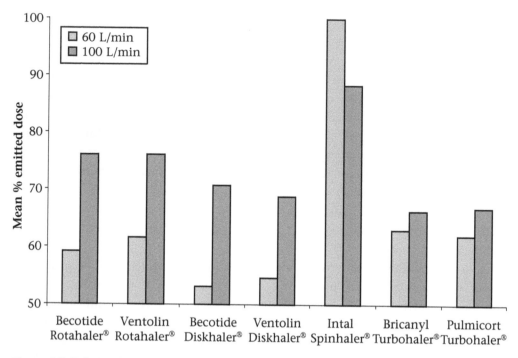

Figure 4.7: Relationship between mean percentage emitted dose of the loaded dose and inhalation flow rate from a range of inhalers (adapted from Hindle *et al.*, 1994).

For example, the resultant aerodynamic diameter of micronised terbutaline sulphate, as measured with a time-of-flight Aerosizer®, was found to be directly dependent upon the shear forces, expressed as the pressure drop across the aerosolising chamber. Increasing the shear forces reduced the aerodynamic diameter until a minimum was obtained at a peak shear force of 4.0 p.s.i. (Hindle and Byron, 1995). The role of flow rate in particle aerosolisation was also clearly demonstrated by an investigation which employed laser light scattering to measure the particle size distribution of micronised salbutamol base from different carrier systems at varying flow rates (MacRitchie *et al.*, 1995) (Table 4.2).

Table 4.2 shows that as the flow rate was increased, the fine particle fraction was increased regardless of the properties of the carrier used. The fine particle fraction is indicative of the deaggregation and dispersion of the powder mixtures. At higher air flow rates, more drug particles were deaggregated and dispersed into the air stream, producing higher fine particle fractions. For example, increasing the flow rate from 28.3 to 80 l min^{-1}, resulted in the fine particle fraction being increased by at least two- to three-fold for all the carriers. The difference in the increase of the fine particle fraction obtained for various carriers may be due to the different adhesion forces which exist between the drug particles and specific carrier.

TABLE 4.2

The fraction of fine particles (<5 μm) of salbutamol base from different carriers at four flow rates (mean % (SD)) (after MacRitchie *et al.*, 1995)

Carriers	Flow rates (l min⁻¹)			
	28.3	40	60	80
Lactose	5.35(1.3)	6.25(1.5)	16.07(3.3)	20.85(1.5)
SD lactose[a]	1.00(0.5)	0.82(2.5)	6.62(3.0)	8.29(2.5)
Sorbitol	6.45(1.8)	8.13(2.4)	25.33(6.4)	32.97(5.7)
Dextrose	2.72(0.9)	2.92(1.1)	5.00(1.7)	7.65(2.2)
Maltose	6.93(1.4)	7.29(1.2)	11.43(3.3)	19.78(5.2)

[a] Spray-dried lactose.

Apart from the velocity of air flow, air turbulence is also important for the entrainment and detachment of drug particles. It is widely accepted that turbulent air flow is more effective than laminar air flow in dispersing the powder mixture (Moren *et al.*, 1985). Turbulence generated within the inhaler device is particularly important for aerosol dispersion. Increasing the flow rate will increase the turbulence of the inhaled air stream, which can be achieved by modifying the design features of dry powder inhaler devices (Timsina *et al.*, 1994).

4.5.2 Particle properties

When compared with the effect of air velocity, particle entrainment and dispersion are dependent upon particle size in a more complex manner. Although particles of different sizes have different aerodynamic properties, the type of variation is mainly governed by the ratio between the weight of the particle and the force of adhesion to either a stationary surface, or the surface of an airborne carrier particle. If the force of adhesion is greater than particle weight, a typical situation for fine particles (e.g. those <20 μm), then increasing the particle size improves particle detachment and dispersion. Thus, for fine particles, the critical velocity required for detachment decreases as particle diameter increases (Soltani *et al.*, 1995). However, when the force of interaction between the contiguous bodies is less than the particle weight, higher drag forces are required to detach or disperse larger particles. Therefore, the critical velocity required for detachment and dispersion will increase with particle size. Within a certain range of particle size, where the adhesion force is commensurate with the particle weight, the critical velocity for detachment to occur is independent of particle size (Zimon, 1982). These relationships between particle size and detachment velocity, V_{det} (the air flow velocity needed to detach 50% adhered particles) were best shown (Figure 4.8) in a study designed to determine the detachment of sylvite dust, from a duct with a diameter of 300 mm and length of 8 m (Zimon, 1982).

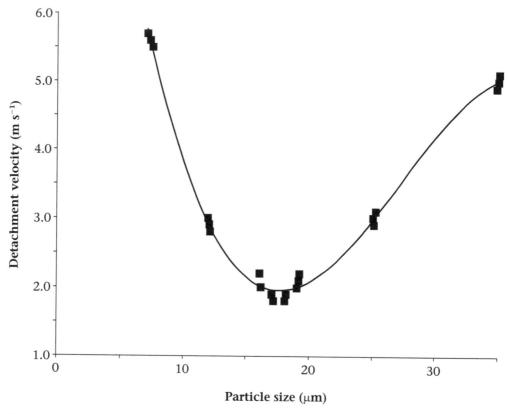

Figure 4.8: Relationship between the size of adhered particles and detachment velocity of an air stream (adapted from Zimon, 1982).

From Figure 4.8, it can be seen that when the particle diameter was less than about 15 μm, under the specific experimental conditions, detachment velocity increased with decreasing particle size. From about 15 to 20 μm, detachment velocity is virtually constant, whereas above 20 μm, an increase in the particle size results in a rise in the detachment velocity (V_{det}) which can be estimated using the following empirical equation (Zimon, 1982):

$$V_{det} = \frac{0.072}{\sqrt{\alpha}} \sqrt{d_p \rho_p + \frac{2F_{ad}}{d_p^2}}$$

(4.18)

where α is a constant of the air duct in kg s^{-1} m^{-1}; d_p is the diameter of the adherent particles (mm) and ρ_p is the density of the powder (kg m^{-3}).

Therefore, due to the relationship between the particle parameters indicated by the terms under the square-root sign in equation (4.18), the detachment velocity may vary either directly or inversely with particle size. For large particles, when F_{ad} is negligible, then the detachment velocity is directly proportional to $d^{1/2}$, i.e. increasing the particle diameter would require a higher velocity air stream to detach a particle from any adhering surface. However, for small particles, when $2F_{ad}/d^2 \gg d_p\rho_p$, the detaching velocity becomes inversely proportional to particle diameter.

A similar relationship between the particle size of fine particles and detachment from a stationary surface by an air stream was also shown in a more recent report (Gotoh *et al.*, 1994a) where latex particles ranging from 1.09 to 11.9 µm were detached from different surfaces. The particle size was shown to be the dominant factor in determining removal efficiency, and the dynamic pressures required to dislodge 50% of the adhered particles from a surface (P_{d50}), increased rapidly with decreasing particle size, regardless of the surface characteristics. For example, P_{d50} for particles of 2.02 µm diameter was more than 10 times higher than that of particles with a diameter of 6.4 µm (Figure 4.9) and increased more markedly with decreasing particle size for smaller than for larger particles with a minimum occurring at about 13–15 µm.

The relationship between the linear flow rate V of the air stream at a specific dynamic pressure was calculated as (Gotoh *et al.*, 1994a):

$$V = \sqrt{\frac{2P_d}{\rho_g}} \tag{4.19}$$

where ρ_g is the air density (kg m^{-3}) and P_d is the dynamic pressure in Pa, and the velocity of 50% removal, V_{50}, was obtained from equation (4.20):

$$V_{50} = \sqrt{\frac{2 \times 1.1 \times 10^5}{d_p^2 \rho_p}} = \frac{470}{d_p\sqrt{\rho_p}} \tag{4.20}$$

where d_p is the diameter of the adhered particles in m.

Thus, the relationship between particle size and detachment velocity predicted by equation (4.20) shows good correlation with that predicted using equation (4.18).

It is impossible from a practical point of view to use these equations to calculate the detachment velocity for particles since it is well known that when cohesive forces begin to affect dispersion behaviour, theories based on the behaviour of a powder of one type will not be applicable to another of a different type (Geldart, 1986). Nevertheless, these equations can provide general guidance for the estimation and comparison of air velocities needed to detach particles of different size under similar conditions.

The removal of pharmaceutical powders by an air stream from a stationary surface, such as the inner walls of capsule shells and inhaler devices, is also dependent upon the size of the particles to be removed. For example, the re-entrainment efficiency of particles of varying sizes from a Spinhaler® capsule was found to be dependent upon particle size (Bell *et al.*, 1971). At the same flow rate, re-entrainment efficiency increased with

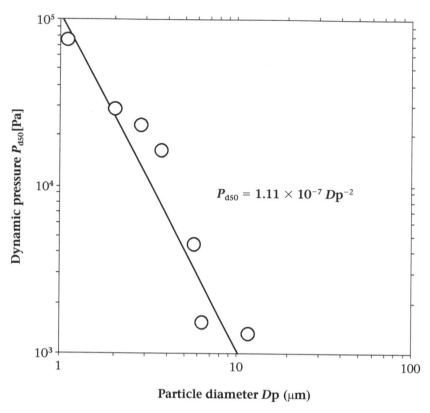

Figure 4.9: Relationship between dynamic pressure P_{d50} and size required to detach latex particles from a glass plate (adapted from Gotoh *et al.*, 1994a).

particle size until it reached a maximum of about 60% at a size of 100 μm. Then, entrainment efficiency decreased drastically with increased particle size and virtually no entrainment was detected when particles were over 120 μm under the experimental conditions employed. The linear velocity (V_i) of an air stream required to initiate entrainment of lactose particles of two large size fractions (90–125 and 125–180 μm) was about 8 m s^{-1}, which was less than half of the V_i (more than 16 m s^{-1}) for two smaller size fractions (63–75 and 75–90 μm) (Staniforth, 1995). The air velocity required to complete entrainment was also lower for the large particles (about 18 m s^{-1}) than for the small ones (about 25 m s^{-1}). The coarser particles became entrained in a smooth continuous manner as discrete particles, whereas the smaller particles exhibited an erratic plug entrainment behaviour, with agglomerates being entrained intermittently.

Although the entrainment of larger particles is relatively easier than that of smaller particles, this does not necessarily indicate that the larger carrier particles are always more favourable than smaller carrier particles for drugs intended to be inhaled. According to equations (4.15) and (4.16), the detachment of fine particles from coarser carriers depends upon the adhesion forces and the mobility of the entrained carrier. Large carrier particles were shown to exert stronger adhesion forces to drugs than smaller ones (Staniforth, 1982). Further, the mobility of airborne smaller particles will be higher than that of the coarser particles. Thus, in terms of drug detachment from airborne carrier particles, smaller ones may be preferred. For example, *in vitro* respirable fractions of salbutamol sulphate from smaller lactose particles were higher than those from larger lactose particles at all flow rates from 60 to 200 l min^{-1} (Ganderton and Kassem, 1992). The increased respirable fraction was indicative of improved detachment of drug particles from the smaller carrier particles. However, the latter often have poorer flowability than larger particles, which may result in other problems during powder handling processes. Therefore, the size of carrier particles has to be optimised according to the entrainment and dispersion properties of the drug and carrier as well as to the flowability of the ordered mix.

4.5.3 Particle–surface interaction

The critical velocity of air flow necessary to detach particles having lower adhesion forces should be less than that needed for particles where these forces are higher. Thus, any factors that affect cohesion and/or adhesion forces would ultimately influence particle detachment.

4.5.3.1 Relative humidity

Environmental relative humidity is an important factor governing particle–particle and particle–surface interactions. The properties and behaviour of solid particulate systems can, for example, be substantially controlled by capillary forces (Schubert, 1984). As mentioned before, relative humidity affects interparticulate forces by means of reducing electrostatic forces and inducing capillary forces. At low relative humidities (<60%) when water vapour condensation on the particles is negligible, increasing relative humidity would reduce interparticulate forces by decreasing electrostatic forces, as a result of improving electrical charge decay. At higher relative humidities when water uptake by the particle is substantial, increasing relative humidity would increase interparticulate forces as a result of capillary forces. Therefore, there may exist an optimum relative humidity where interparticulate forces are at a minimum and the particle removal is optimised. For example, increasing the ambient relative humidity was found to increase the removal efficiency of latex (styrene/divinylbenzene) particles of 3.7 μm diameter from a glass surface by air streams at relative humidities up to 67%, when particle removal reached a peak value (Gotoh *et al.*, 1994b). As the relative humidity was increased further so the removal efficiency of particles was decreased (Figure 4.10). A similar tendency was also observed for particle removal from other surfaces, such as

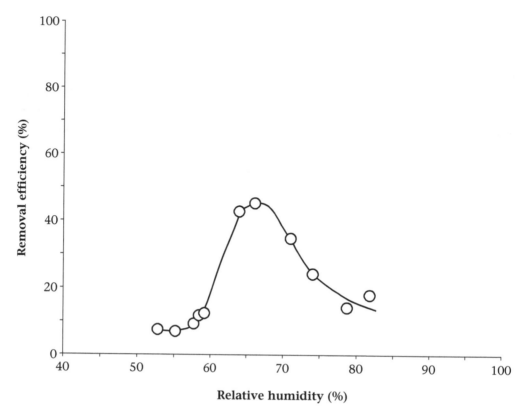

Figure 4.10: Effects of relative humidity on removal of styrene/divinylbenzene particles from a glass plate (adapted from Gotoh *et al.*, 1994b).

metals and plastics (Gotoh *et al.*, 1994c). The humidity for maximum removal efficiency was dependent upon the surface roughness of the substrate and its increase required a higher relative humidity for maximal removal of particles.

If sufficient water vapour is allowed to condense on the surface of particles, especially during storage and/or fluidisation at high relative humidities (>60%), interparticulate forces due to capillarity may reduce not only fluidisability but also the flowability of powders. The addition of small quantities of a non-volatile liquid to a group B powder (Table 4.1) caused its fluidisation behaviour to move through group A to group C (Seville and Clift, 1984). Most group A powders can be made to behave as powders from group C by fluidising them with air of a relative humidity within the range of 60–90% (Geldart *et al.*, 1984).

Similar effects of relative humidity on the entrainment of dry powders for inhalation were also observed (Jashnani *et al.*, 1995). When micronised salbutamol base (1.36 µm) and sulphate (2.5 µm) particles were emitted from a model dry powder inhaler under varying environmental relative humidity and temperature into a twin-stage liquid impinger, both the amount emitted from the mouthpiece of the inhaler and the fine particle fraction that reached stage 2 of the impinger, as expressed as a fraction of either the loaded dose or the emitted dose, were found to decrease with increasing relative humidity at any given temperature, with the difference being more marked at higher temperatures. Under conditions of elevated relative humidity and temperature, the interparticulate forces (cohesive forces) between the fine particles would have been increased due to capillary forces, resulting in poorer fluidisation or entrainment of the particles into an inhaled air stream and leading to a decreased emitted dose from the inhaler. After entrainment, the drag forces of the inhaled air stream were probably not sufficient to deaggregate particle agglomerates and hence the percentage of fine particles would be reduced.

The effect of relative humidity on particle behaviour may be more pronounced for small particles than for large particles since a liquid layer is more likely to form on the surface of the former. Thus, relative humidity may have greater effects on powder formulations employing only fine drug powders in comparison to those composed of fine drug particles mixed with coarse carrier particles. For example, the performance of dry powders composed of drug and carrier particles from the Ventolin Rotahaler®, was largely unaffected by relatively extreme test conditions (i.e. equilibration at 45°C and 80% relative humidity for 30 min), with respect to both the fraction of fine particles and the emitted dose when tested with a twin-stage liquid impinger (Hindle et al., 1994).

Although capillary forces appear to be decisive in determining the removal efficiency of some particles, especially those of hydrophilic materials, condensation of water vapour onto the particles is a function of the duration of their exposure to elevated environmental humidities. If the powders are exposed to the environment for such a short period of time that water vapour condensation does not become manifest, even highly hygroscopic particles may not be substantially influenced by high humidities. This may be the main reason that the performance of dry powders composed of drug and carrier particles from the Ventolin Rotahaler® was largely unaffected by the extreme test conditions described above (Hindle et al., 1994). However, if water vapour is allowed to condense on particle surfaces for a prolonged period of time, especially at elevated temperatures, it will not only increase interparticulate forces but will also induce growth of drug particles due to migration of adsorbed water molecules into those particles. Both effects will reduce the delivery efficiency of drugs (Jashnani et al., 1995). For example, both the emitted dose and the fine particle fraction of salbutamol base (1.36 µm) and sulphate (2.5 µm) particles from a model dry powder inhaler, were found to decrease with increasing relative humidity at any given temperature, with the difference being more marked at higher temperatures (Jashnani et al., 1995).

CHAPTER 4

4.5.3.2 Contact area

A smaller contact area (r_c) between particles and any surface will lead to easier detachment of particles. For example, mica particles consistently required higher air velocities to become detached from a microscope slide than glass spheres of a similar size although the former showed lower adhesion forces than the latter, as measured by a centrifugation method (Davies, 1962). This was shown to be due to a greater contact area between mica particles and the surface because of their plate-like shape. According to a more recent calculation (Soltani and Ahmadi, 1995), the critical shear velocity required to detach rubber particles from both smooth and rough surfaces was at least 10 times larger than that for graphite particles because, on contact, the former are more easily deformed than the latter. Rubber particles will therefore have a larger contact area with a surface and hence will exhibit a higher frictional coefficient as compared with graphite particles under similar adhesive conditions. Furthermore, as mentioned above, since adhesion forces are a function of contact area, the higher this is the greater the adhesion force. Any such increase due to a larger contact area will further increase the velocity of air needed to dislodge the particles.

The contact area also depends on the applied pressure, hardness of both the particle and the surface as well as the orientation of the particles (if they are irregularly-shaped). For example, plate-like particles will have the largest contact area with a substrate surface when adhered in their most stable orientation, i.e. with the largest plane normal to the substrate surface. Meanwhile, drag forces acting on the particle will be lowest since the area of the particle perpendicular to the direction of the flow is the smallest. Thus, irregularly-shaped particles would be less likely to roll than more spherical particles under similar flow conditions.

4.5.3.3 Surface smoothness of carrier

Increasing surface smoothness usually increases adhesion forces between particles as a result of an increased contact area between the interacting species. Unless mechanical interlocking is implicated in adhesion, decreasing carrier surface smoothness would improve detachment of adhered particles. For example, surface smoothness of the substrate was shown to affect the detachment of latex particles by an air stream (Gotoh *et al.*, 1994a). Surface smoothness was expressed as the inverse of an average radius of curvature of surface protrusions (r_w), a higher r_w^{-1} indicating higher surface roughness. The dynamic pressures of compressed air required to detach 50% of the adhered latex particles (P_{d50}) of different sizes from surfaces of varying smoothness are shown in Figure 4.11. It can be seen that P_{d50} decreased with increasing surface roughness, indicating that the particles adhered to a rougher surface are more easily detached than similar particles adhered to a smoother one. This was attributed by the authors to decreased van der Waals forces between particles and rougher surfaces (Gotoh *et al.*, 1994a). However,

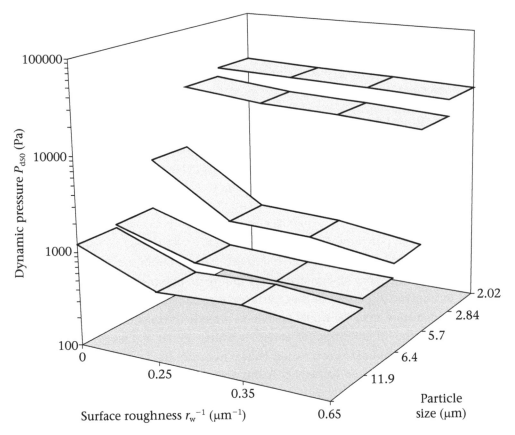

Figure 4.11: Effects of particle size and surface smoothness on the dynamic pressure of an air stream required to detach 50% of adhered particles from surfaces of varying smoothness (adapted from Gotoh *et al.*, 1994a).

the change in the near-wall fluid flow properties due to variation in the surface smoothness may also contribute partly to the effects. Nevertheless, particle size had more pronounced effects on detachment than surface smoothness under specific conditions. Smaller particles were more difficult (higher P_{d50}) to detach than larger ones regardless of the surface materials used.

The detachment of particles having different hardness from surfaces of varying smoothness has been modelled theoretically, based upon particle rolling, sliding and lifting mechanisms (Soltani *et al.*, 1995). Increasing the surface roughness of the substrate has been shown to reduce the critical acceleration required to detach all the particles investigated. This was attributed to reduced adhesion forces of the particles to the surface due to the decreased contact area at high surface roughness.

Increasing surface roughness may further improve particle detachment by changing the fluid properties in the vicinity of adhered particles to be detached. The near-wall flow behaviour is expected to have a profound effect on particle detachment where turbulent flow is thought to be dominated by vertical coherent structures and occasional bursts (Hinze, 1975). Increasing surface roughness would increase the turbulence of flow near the particles to be detached and, consequently, would improve particle detachment at similar velocities.

4.5.3.4 Frictional coefficient

Although increasing surface roughness may improve particle detachment, this effect may be reduced by an increased frictional coefficient between particles and surfaces at higher surface roughness. According to equations (4.12) to (4.16), decreasing the frictional coefficient (μ) between fine particles and coarser carriers would favour detachment. The flowability of lactose particles was found to be greatly increased by the coating with certain polymers (Fernandez-Arevalo et al., 1990), due to a reduction in the interparticulate friction as a result of improved surface smoothness. The coating of carrier particles with other biodegradable and non-toxic polymers may be a reasonable proposal to improve dispersion of drug particles, especially for highly expensive drugs. An optimised coating material would be able to provide a reduction in interparticulate forces through the modification of surface texture, electrical charge and moisture absorption. Such combined effects would reduce not only the adhesion forces but also the frictional coefficient and consequently improve the dispersion of drug particles on inhalation.

4.5.4 Exposure time of particles to the air stream

The effects of exposure time of the particles to the air stream on the efficiency of particle removal is not negligible. If the time of exposure of particles to an air stream is increased, a higher degree of particle removal will be achieved (Corn, 1966). An empirical equation (equation (4.21)) was generated to estimate the removal efficiency, $\eta(t)$, at different jet duration times:

$$\eta(t) = \eta_{max}\left[1 - \exp\left(-\frac{t}{\tau}\right)\right] \tag{4.21}$$

where η_{max} is the maximum removal efficiency under the conditions and τ is a time constant.

The removal of latex (styrene/divinylbenzene) particles (3.7 μm) from a glass surface by a compressed air stream was shown to reach equilibrium within 2 s (Gotoh et al., 1995). Since it usually takes less than 1 s for most inhaled particles to impact on a surface, this short period of time may not be sufficient for complete detachment of drug particles from the carrier. Thus, any means of increasing the exposure time would result in more drug detachment. For example, extension devices, such as those used in MDIs to aid evaporation of the propellant, may also be beneficial for delivery from dry powder inhalers although the volume of any such device might need to be minimised.

In conclusion, particle interaction with a fluid is a function of both the fluid properties and particle characteristics. Increasing the fluid velocity or the size of the interactive particle will result in an increase in the drag force of the fluid acting on the particle. Such a force is crucial in determining whether a powder is fluidised or dispersed when it is exposed to the fluid. If the drag force overcomes the combined effects of gravitational and interparticulate forces, then the powder is likely to be dispersed in the fluid. Removal of adhered particles from either a stationary surface or airborne carrier particles is dependent upon many factors such as the drag forces of the air stream acting on the particles, the size of both the adhered and carrier particles and the adhesional and frictional forces between the adhered particles and substrate surface. All these factors have to be carefully controlled with a view to improving the removal efficiency of any adhered particles from the substrate surfaces.

REFERENCES

ALLEN, M.D. and RAABE, O.G. (1985) Slip correction measurements of spherical solid aerosol particles in an improved Millikan apparatus. *Aerosol Sci. Tech.* **4**, 269–286.

BARON, P.A. and WILLEKE, K. (1993) Gas and particle motion. In: WILLEKE, K. and BARON, P.A. (eds), *Aerosol Measurement: Principles, Techniques, and Applications.* New York, Van Nostrand Reinhold, 23–40.

BELL, J.H., HARTLEY, P.S. and COX, J.S.G. (1971) Dry powder aerosols. I. A new powder inhalation device. *J. Pharm. Sci.* **60**, 1559–1564.

BUCKTON, G. (1995) *Interfacial Phenomena in Drug Delivery and Targeting.* Chichester, Harwood Academic Publishers, 27–58.

CLIFT, R., GRACE, J.R. and WEER, M.E. (1978) *Bubbles, Drops and Particles.* New York, Academic Press, 142.

CORN, M. (1966) Adhesion of particles. In: DAVIES, C.N. (ed.), *Aerosol Science.* London, Academic Press, 359–392.

DAVIES, C.N. (1962) Measurement of particles. *Nature (Lond.)* **195**, 768–770.

ENDO, Y., HASEBE, S. and KOUSAKA, Y. (1997) Dispersion of aggregates of fine powder by acceleration in an air stream and its application to the evaluation of adhesion between particles. *Powder Technol.* **91**, 25–30.

FERNANDEZ-AREVALO, M., VELA, M.T. and RABASCO, A.M. (1990) Rheological study of lactose coated with acrylic resins. *Drug Dev. Ind. Pharm.* **16**, 295–313.

CHAPTER 4

FRIEDLANDER, S.K. (1977) *Smoke, Dust and Haze: Fundamentals of aerosol behavior.* New York, Wiley, 105.

GANDERTON, D. (1992) The generation of respirable cloud from coarse powder aggregates. *J. Biopharm. Sci.* **3**, 101–105.

GANDERTON, D. and KASSEM, N.M. (1992) Dry powder inhalers. In: GANDERTON, D. and JONES, T. (eds), *Advances in Pharmaceutical Sciences*, Vol. 6. London, Academic Press, 165–191.

GELDART, D. (1972) The effect of particle size and size distribution on the behaviour of gas-fluidised beds. *Powder Technol.* **6**, 201–215.

GELDART D. (1986) *Gas Fluidization Technology.* Chichester, Wiley.

GELDART, D., HARNBY, N. and WONG, A.C. (1984) Fluidization of cohesive powders. *Powder Technol.* **37**, 25–37.

GOTOH, K., KIDA, M. and MASUDA, H. (1994a) Effect of particle diameter on removal of surface particles using high speed air jet. *Kagaku Kogaku Ronbun.* **20**, 693–700.

GOTOH, K., TAKEBE, S., MASUDA, H. and BANBA, Y. (1994b) The effect of humidity on the removal of fine particles on a solid surface using a high-speed air-jet. *Kagaku Kogaku Ronbun.* **20**, 205–212.

GOTOH, K., TAGAYA, M. and MASUDA, H. (1995) Mechanism of air–jet removal of particles. *Kagaku Kogaku Ronbun.* **21**, 723–731.

GOTOH, K., TAKEBE, S. and MASUDA, H. (1994c) Effect of surface material on particle removal using high speed air jet. *Kagaku Kogaku Ronbun.* **20**, 685–692.

HAIDER, A. and LEVENSPEIL, O. (1989) Drag coefficient and terminal velocity of spherical and nonspherical particles. *Powder Technol.* **58**, 63–70.

HAMMOND, R. (1958) *Dispersion of Materials.* New York, Philosophical Library.

HEYWOOD, H. (1962) In: *Proceedings of the Symposium on Interaction in Particles and Fluids*, Vol. 189. London, Institute Chemical Engineering, 226, 231, 241, 243.

HINDLE, M. and BYRON, P.R. (1995) Size distribution control of raw materials for dry-powder inhalers using the aerosizer with the aero-disperser. *Pharm. Technol.* June, 64–78.

HINDLE, M., JASHNANI, R.N. and BYRON, P.R. (1994) Dose emissions from marketed inhalers: influence of flow, volume and environment. *Respiratory Drug Delivery* **IV**, 137–142.

HINZE, J.O. (1975) *Turbulence*, 2nd edn. New York, McGraw-Hill.

JASHNANI, R.N., BYRON, P.R. and DALBY, R.N. (1995) Testing of dry powder aerosol formulations in different environmental conditions. *Int. J. Pharm.* **113**, 123–130.

KOUSAKA, Y., IN LINOYA, K., GOTOH, K. and HIGASHITANI, T. (eds) (1991) *Powder Technology Handbook.* New York, Marcel Dekker, 417.

MACRITCHIE, H.B., MARTIN, G.P., MARRIOTT, C. and MURPHY, L. (1995) Determination of the inspirable properties of dry powder aerosols using laser light scattering. *J. Pharm. Pharmacol.* **47**, 1072.

MATSUSAKA, S. and MASUDA, H. (1996) Particle reentrainment from a fine powder layer in a turbulent air flow. *Aerosol Sci. Tech.* **24**, 69–84.

MOREN, F., NEWHOUSE, M.T. and DOLOVICH, M.B. (1985) *Aerosol in Medicine: Principles, Diagnosis and Therapy.* Amsterdam, Elsevier.

PATEL, N.K., KENNON, L. and LEVINSON, R.S. (1986) Pharmaceutical suspensions. In: LACHMAN, L., LIEBERMAN, H.A. and KANIG, J.L (eds), *The Theory and Practice of Industrial Pharmacy,* 3rd edn. Philadelphia, Lea & Febiger, 479–533.

RADER, D.J. (1990) Momentum slip correction factor for small particles in nine common gases. *J. Aerosol Sci.* **21**, 161–168.

RANKELL, A.S., LIEBERMAN, H.A. and SCHIFFMAN, R.F. (1986) Drying. In: LACHMAN, L., LIEBERMAN, H.A. and KANIG, J.L (eds), *The Theory and Practice of Industrial Pharmacy,* 3rd edn. Philadelphia, Lea & Febiger, 47–65.

SARTOR, J.D. and ABBOTT, C.E. (1973) Prediction and measurement of the accelerated motion of water drops in air. *J. Appl. Meteorol.* **14**, 232–239.

SCHUBERT, H. (1984) Capillary forces–Modelling and application in particulate technology. *Powder Technol.* **37**, 105–116.

SEVILLE, J.P.K. and CLIFT, R. (1984) The effect of thin liquid layers on fluidisation characteristics. *Powder Technol.* **37**, 117–129.

SOLTANI, M. and AHMADI, G. (1995) Particle detachment from rough surfaces in turbulent flows. *J. Adhesion* **51**, 105–123.

SOLTANI, M., AHMADI, G., BAYER, R.G. and GAYNES, M.A. (1995) Particle detachment mechanisms from rough surfaces under substrate acceleration. *J. Adhes. Sci. Technol.* **9**, 453–473.

STANIFORTH, J.N. (1982) Relationship between vibration produced during powder handling and segregation of pharmaceutical powder mixes. *Int. J. Pharm.* **12**, 199–207.

■ CHAPTER 4 ■

STANIFORTH, J.N. (1995) Performance-modifying influences in dry powder inhalation system. *Aerosol Sci. Tech.* **22**, 346–353.

TIMSINA, M.P., MARTIN, G.P., MARRIOTT, C., GANDERTON, D. and YIANNESKIS, M. (1994) Drug delivery to the respiratory tract using dry powder inhalers. *Int. J. Pharm.* **101**, 1–13.

ZIMON, A.D. (1982) Adhesion, molecular interaction and surface roughness. In: *Adhesion of Dust and Powders*, 2nd edn. London, Consultant Bureau, 46–47.

Particulate Interactions in Dry Powder Aerosols

5

Contents

5.1   INTRODUCTION

All dry powder inhalers (DPIs) have four basic features: (a) a reservoir and/or a dose-metering mechanism; (b) an aerosolisation mechanism; (c) a deaggregation mechanism and (d) an adapter to direct the aerosol into the patient's mouth. The major components of most DPIs are micronised drug powders adhered to coarse carrier particles, a drug reservoir or premetered individual dose, the body of the device, and a cover to prevent ingress of dust or moisture. In order for the drug to gain access to the lower airways, it is generally accepted that a prime requirement is that the drug particles have an aerodynamic diameter between 1 and 5 µm (Newman and Clark, 1983; Gonda, 1990). An important consequence of this requirement for fine particles for inhalation arises from the fact that powder flow properties are dependent on the particle size distribution, fine particles generally flowing less well than coarse ones. The final formulation must flow sufficiently well either to be dispensed from the bulk reservoir to give an adequately reproducible dose, or be capable of being handled well on automatic filling machines to produce the unit dose forms for use in a device. Small particles are also notoriously difficult to disperse (Hickey *et al.*, 1994) due to their highly cohesive nature. Fine drug particles must therefore be formulated to have appropriate properties such as reasonable flowability and high dispersibility. Two principal approaches have been employed to improve the flow and dispersion characteristics of drug particles. The first involves the controlled aggregation of the undiluted drug to form loosely adherent floccules (Bell *et al.*, 1971); such an approach takes advantage of the inherent cohesiveness of the drug particles. The alternative approach is to use formulations comprised of fine drug particles blended with coarser carrier particles (Ganderton, 1992).

Three major processes are involved in the delivery of drugs from DPIs, namely, the detachment of drug particles from the carrier, their dispersion in the air flow and deposition in the respiratory tract. Thus, any factor that affects any of these processes could ultimately influence the bioavailability of the inhaled drug. These include the design of inhaler devices, airway physiology, inhalation manoeuvres and the powder formulation (Table 5.1). For example, the drug–carrier particle interaction (adhesion) will greatly influence the detachment of drug particles from their carrier whilst the drug–drug interaction (cohesion) will affect the deaggregation of drug particles into single particles suitable for deep lung penetration. In addition, the design of the inhaler device, inhalation flow rate and turbulence of the inhaled air stream will also determine the efficiency of particle–particle separation. Furthermore, the particle size, shape and density are all crucial determinants of the aerodynamic properties of the particles, which in turn determine particle dispersion in an air stream and eventually deposition of the drug in the respiratory tract. Moreover, the physiology of the airways and breathing patterns including tidal volume, breathing frequency and breath-holding, etc., are also factors governing the deposition profiles of the dispersed, individual particles.

TABLE 5.1

Some of the major factors affecting drug delivery from DPIs

Design of inhaler devices	Rotahaler®, Spinhaler®, Inhalator®, Turbohaler®, Diskhaler®, Diskus®, Accuhaler®, etc.
Powder formulation	Drug–drug interaction, drug–carrier interaction, particle size and shape, surface texture, crystallinity, hygroscopicity, density etc.
Inhalation manoeuvre	Flow rate, tidal volume, breath-holding, breathing frequency, air acceleration, etc.
Physiology of respiratory tract	Geometry, lung function, pulmonary ventilation, mucus secretion, etc.

Numerous reports and review articles have been published describing the influence of these various factors on the overall delivery efficiency (often determined by the fraction of the total dose that can reach the lower airways) of drugs from DPIs (see Hickey, 1992). However, most currently used DPIs fail to possess a high delivery efficiency (Timsina *et al.*, 1994) since only approximately 10% of the total administered dose reaches the lower airways. About 10% total dose is retained in the inhaler devices and over 80% is found to deposit in the upper airways, most of which is eventually swallowed (Gupta and Hickey, 1991). Thus, the optimisation of drug delivery from DPIs to the lower airways with the hope of increasing therapeutic effect and decreasing any possible side-effects, has been a major concern in the development of DPIs. However, the bulk of the previous work has focused on inhaler device design and surprisingly, little has been published on approaches to improve the delivery efficiency of dry powder formulations. It is a truism that the formulation factors of dry powders are equally, if not more, important than the design of inhaler devices in the optimisation of drug delivery to the lung. For example, micronised drug particles are usually present in low concentrations in the powder formulation, with a drug to carrier ratio of $1:67.5$ being typical. A large portion of drug particles may therefore be expected to be adhered to potential binding sites of the carrier particles or entrapped in any surface crevices existing on the carrier surfaces. The resultant strong interaction of drug with carrier particles impedes drug detachment and dispersion in the inhalation air stream and, consequently reduces the overall deposition of drug particles in the respiratory tract (Figure 5.1). Insufficient detachment of drug from the carrier due to strong interparticulate forces may be one of the major causes of low delivery efficiency encountered with most DPIs. Therefore, increasing drug detachment and dispersion by means of reducing interparticulate forces in powder formulations may provide one of the major strategies to provide efficient drug delivery to the lower airways.

Since particle–particle interaction is primarily a surface phenomenon, it is mainly dependent upon the surface texture, particle size and shape of the interactive particles. Increasing surface smoothness of lactose carrier particles was shown to increase the

CHAPTER 5

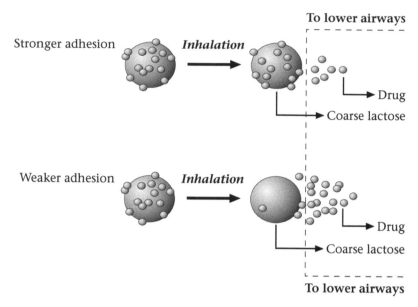

Figure 5.1: Relationship between adhesion, detachment and lung penetration of drug particles.

deposition of salbutamol sulphate from dry powder inhalers, the fraction of fine drug particles <5 μm being increased from 4.1 to 23%, after aerosolisation at 60 l min^{-1} via a Rotahaler®, when the surface rugosity of lactose particles was reduced (Ganderton and Kassem, 1992). This was attributed to a reduced force of interaction of the drug with the carrier particles due to the increased surface smoothness of the carrier. However, adhesion of fine particles to coarse particles is also determined by other factors such as particle size, the shape of both components, their relative concentrations, the existence of amorphous regions and active binding sites on the particle surface, addition of ternary components, etc. The control of any individual factor such as surface smoothness would appear to be insufficient to minimise drug–carrier interactions and maximise detachment of drug particles from carrier particles on inhalation. The combination of controlling these various factors mentioned above and in earlier chapters would provide a more extensive strategy to optimise drug delivery from DPIs.

5.2 GENERATION OF AEROSOL DRY POWDERS

Dry powders for inhalation are mostly produced by mechanical micronisation although spray-drying has been investigated as an alternative means of generating particles suitable for deep lung penetration (Vidgren *et al.*, 1987). More recently, recrystallisation from supercritical solution has been suggested to represent a further potential method (Phillips and Stella, 1993). Powders produced using different methods would be expected to have different physical, if not chemical, properties and these, as has been stated, include particle size, size distribution, shape, crystalline properties, surface texture and energy, etc. These physical properties are critical in the behaviour of particles, both before and after inhalation.

5.2.1 Micronisation

Attrition milling has long been used as the main method to prepare inspirable particles. Opposing jets of compressed air at different pressures force the particles to impact with one another and this violent attrition causes the reduction in particle size. A cyclone separation mechanism is used to collect the resultant micronised powder. To obtain particles that can reach the lower airways, large quantities of energy are needed during micronisation and this energy can affect the material being processed (Florence and Salole, 1976; Otsuka and Kaneniwa, 1990). Thus, micronised particles often have both a high surface energy and electrical charge (Buckton *et al.*, 1988). Moreover, crystalline disorder can be introduced onto particle surfaces during mechanical micronisation and if this disorder is more significant than the occasional molecular dislocation, the disorder induced can be regarded as an amorphous region (Saleki-Gerhardt *et al.*, 1994). These amorphous regions are thermodynamically unstable and will convert to more stable crystal forms under appropriate environmental conditions such as at elevated temperatures and/or relative humidities. Ward and Schultz (1995) have investigated the changes in crystallinity of salbutamol sulphate, a β_2-agonist which has been widely employed as a bronchodilator by inhalation as a dry powder, during mechanical micronisation and the possible implications for its inhaled delivery. After micronisation, the salbutamol sulphate powder showed an extra, small exothermic peak at 85°C in the differential scanning calorimetry (DSC) thermogram, due to crystallisation of the amorphous drug. However, after 24 h storage at 40°C/75% relative humidity, the crystallisation exotherm of the micronised particles disappeared, suggesting that the amorphous region introduced during the micronisation process had converted to a crystalline state under such conditions. The morphology of micronised particles was found to have changed distinctively after storage, the crystal edges becoming more rounded and smoother when compared to the same micronised particles before storage. Significant interparticulate bridging was also observed between the stored particles and the mass median diameter of the micronised particles, measured by laser light diffraction, increased from 1.5 μm before storage to 3.9 μm after 24 h at 40°C/75% RH. These results are indicative of the importance of the effects of crystallinity disorder, however small it may be in terms of total surface area, on the bulk properties of the micronised powders. In an earlier report, Ahlneck and Zografi (1990) showed that if particles have a small proportion of amorphous regions on the surface, then it is likely that water vapour adsorbed will be concentrated on these areas. Thus, the insignificant amount of water relative to the bulk of the powder, will in fact represent a large amount of water at the amorphous regions. The effects of water may thus be amplified at the amorphous regions on the particle surface and even small amounts can therefore produce marked changes in the molecular mobility in disordered regions, resulting in a significant change in physical and even chemical (Konno, 1990) properties of the bulk powder. The surface disorders on particles induced by micronisation may be even more significant with dry powders intended for inhalation than those prepared for other pharmaceutical purposes since both the *in vitro* and *in vivo* behaviour of dry powder formulations are critically dependent upon physical properties.

Apart from introducing amorphous regions onto particle surfaces, attrition milling may induce physico-chemical instability of some drugs although it does not expose drugs to extremely high temperature. For example, attrition milling is probably not suitable for drugs that are sensitive to compaction since violent collisions among drug particles may have a similar effect on the drug to that encountered during compaction. The introduction of crystalline disorder to the chemical stability of macromolecules such as proteins and peptides may be even more considerable than in the case of smaller molecules since the reconversion of the amorphous regions to the original crystalline state could induce serious chemical instability reactions such as the breakdown of the three-dimensional and conformational structure of a protein drug (Briggner, 1994). Furthermore, during attrition milling, excessively high temperatures may still be induced on the points where collision occurs, which may lead to destabilisation of the drugs in these regions. Therefore, micronisation of these materials should be carried out under carefully controlled conditions, such as low temperatures. Since respirable particles require a size of less than 5 μm, then a vigorous micronisation process is imperative to produce such fine particles. The conditions under which the particles are processed have been shown to influence the surface energy of the final products (Buckton *et al.*, 1988) and it is reasonable to assume that any such difference in the processing conditions might also induce different surface disorder onto micronised materials. Apart from particle size, size distribution, shape and other conventionally employed standards, the micronisation process for the production of inhaled particles should also be rigorously controlled to ensure maximum surface crystallinity of the final product. Since the formation of crystal disorder is unavoidable if mechanical micronisation is employed, it may be necessary to carry out treatment of the particles after micronisation in a controlled manner to deliberately convert any unstable amorphous regions. Such a transformation would greatly decrease the sensitivity of powders to any surrounding environment and thus enhance uniformity and reproducibility of *in vitro* and *in vivo* performances of the drug particles.

5.2.2 Spray-drying

Spray-drying has been employed as a routine technique for the production of pharmaceutical particles for decades. Spray-drying constitutes a single step process that transforms a solution or suspension into a fine powder. The feed material in spray-drying is a liquid, and drying is accomplished by atomising this into a hot drying medium. Spray-drying involves droplet formation and rapid evaporation of the solvent. The properties of spray-dried particles are affected by many factors such as initial drug concentration, feed rates and temperatures, etc. Spray-drying is a potential alternative means of generating fine particles suitable for inhalation and it has also been employed to manufacture fine particles of peptidic formulations (Mumenthaler *et al.*, 1991). Generally, spray-drying produces hollow spherical particles, resulting in a powder with low bulk density compared to the initial material. However, particles manufactured in this manner demonstrate poor flow characteristics. Furthermore, the need to provide heat during particle formation by spray-drying makes it less suitable for thermosensitive compounds.

Due to the rapid evaporation of the solvent, spray-dried particles are often composed of amorphous forms of drugs and are usually more spherical in shape than micronised particles (Vidgren *et al.*, 1987). In the latter study, some of the physical properties of spray-dried disodium cromoglycate (DSCG), prepared from an ethanol/water mixture, were compared with those of the DSCG produced by mechanical micronisation. The spray-dried particles, with an arithmetic mean diameter of 2.81 μm, were found to possess a higher respirable fraction (particles <3.3 μm, 26.4%) than micronised particles with an arithmetic mean diameter of 3.81 μm (particles <3.3 μm, 10.1%). However, it was unspecified whether the improved respirable fraction for spray-dried particles was due to an improvement in the aerodynamic properties or the reported lower geometric diameter of the spray-dried particles. Similar results were also obtained when DSCG particles were administered via an MDI. Spray-dried drug was reported to have over twice the respirable fraction (11.9%) of that of the micronised particles (5.0%) (Vidgren *et al.*, 1988). However, spray-dried DSCG was found to be amorphous, without distinctive peaks on the X-ray diffraction pattern, but after storing at 60% relative humidity, the crystallinity of the particles was markedly increased, as shown by an increased number of peaks on the X-ray diffraction patterns (Vidgren *et al.*, 1988). Scanning electron micrographs also showed that some non-spherical, typically crystalline particles appeared in spray-dried powders after storage at 60% relative humidities but no changes in particle shape were observed when the particles were stored at lower relative humidities. The recrystallisation of amorphous DSCG from spray-dried particles could also be detected from the moisture adsorption and desorption isotherms. The spray-dried particles were shown to take up more than 30% water when stored at relative humidities over 60% whereas the micronised particles only adsorbed about 15% water under

■ CHAPTER 5 ■

similar conditions. Thus, the fraction less than 7.1 μm for spray-dried particles was markedly reduced from about 30% when stored at 0% RH to less than 10% when stored at about 80% RH. However, the fractions of micronised particles were reported to be more or less constant at relative humidities from 0 to 80%.

Chawla *et al.* (1994) were able to prepare spray-dried salbutamol sulphate particles with a mass median diameter of 4.5 μm. The particles were shown to be spherical in shape and the drug was mostly in an amorphous form. After mixing with median grade lactose (MMD 52.6 μm) in concentrations of 0–5%, both micronised and spray-dried particles slightly increased the flowability of lactose at drug particle concentrations of less than about 1%. Further increases in the concentration of drug particles resulted in a decrease in the flowability of lactose. At all the drug particle concentrations, blends of micronised particles showed higher flowabilities than those of spray-dried particles. The percentages of the respirable fractions for powdered aerosols liberated from a Spinhaler®, as measured by a twin-stage liquid impinger, were 7.63 ± 3.02% and 10.36 ± 6.76% for spray-dried and micronised salbutamol sulphate, respectively. These results were confirmed by a more recent report where three different inertial impaction methods were employed to compare the *in vitro* deposition of micronised and spray-dried salbutamol sulphate from a Rotahaler® (Venthoye *et al.*, 1995). The mean percentage respirable mass varied from 7.01 to 14.42% for micronised particles, which was markedly higher than that obtained (2.81–4.76%) using spray-dried salbutamol sulphate. Thus, spray-dried particles produced fewer inspirable particles after inhalation as compared with micronised particles, although they were prepared to have smaller geometric diameters and narrower size than the micronised particles (Venthoye *et al.*, 1995). In spite of the spherical shape of spray-dried particles, they were also shown to be less flowable than the more anisometric micronised particles (Chawla *et al.*, 1994). The reduction in flowability and respirable fraction of the spray-dried drug particles may be partly due to the stronger interparticulate forces, brought about as a result of greater deformation at contact sites due to the hollow and amorphous nature of spray-dried particles. Moreover, as mentioned before, the amorphous content will render the spray-dried particles more sensitive to change in the RH of the surrounding atmosphere than micronised particles. Higher water uptake during storage and the associated transformation of amorphous to more stable crystalline forms will undoubtedly result in a significant increase in interparticulate forces and change in the particle characteristics such as size and shape. The higher hygroscopicity of spray-dried particles may also lead to a significant growth in size of inhaled particles in the humid airways. These disadvantages may markedly reduce some of the advantages of employing spray-dried particles as drug powders for inhalation as observed in earlier studies (e.g. Vidgren *et al.*, 1987).

More recently, spray-drying has been employed to prepare protein powders for inhalation (Oeswein and Patton, 1990; Mumenthaler *et al.*, 1994; Chan *et al.*, 1997). Although lyophilisation (freeze-drying) is a widely employed technique to prepare powdered proteins and peptides, this method is an energy-intensive and time-consuming process. Moreover, freeze-drying often results in a broad particle size distribution and the dried materials have to undergo a subsequent treatment, such as micronisation, in order to obtain inspirable particles. Partial denaturation of these thermosensitive compounds may occur during mechanical micronisation because of the high localised temperature induced by the violent collisions and friction between the micronised particles. These problems associated with freeze-drying may be solved by using spray-drying since it comprises of a single step and is particularly economical when compared with freeze-drying. Most importantly, spray-drying is able to produce particles of a size suitable for inhalation and thus, spray-dried particles may be directly formulated into dry powder aerosols. The denaturation of proteins during spray-drying may be minimised by carefully controlling the operational parameters. For example, approximately 25% of recombinant methionyl human growth hormone (hGH) was found to be degraded during spray-drying at 90–150°C but tissue-type plasminogen activator remained intact under similar conditions (Mumenthaler *et al.*, 1994). The addition of 0.1% (w/v) polysorbate 20 to hGH solutions was shown to reduce the destabilisation of the protein upon atomisation, which was found to occur mainly at the air–liquid interface (Oeswein and Patton, 1990); polysorbate 20 may have replaced hGH molecules at the air–liquid interface of droplets in the spray and thus reduced exposure of the protein to surface denaturation. In another study, recombinant human deoxyribonuclease (rhDNase) was spray-dried either alone or with sodium chloride and the dried particles were mixed with different coarse carriers such as lactose, mannitol and sodium chloride (Chan *et al.*, 1997). rhDNase was found to remain stable after spray-drying at an inlet temperature of 90°C, which produced pure rhDNase powder with a fine particle fraction (FPF; % w/w of particles <7 μm) of about 20%. The dispersibility of the powder was improved after sodium chloride was included and there was a linear relationship between the NaCl content and FPF for a similar primary size (~3 μm volume median diameter) of particle. This was attributed by the authors to the higher crystallinity of the powders containing higher NaCl content (Chan *et al.*, 1997). After blending with the carrier particles, an overall two-fold increase in the FPF of the drug in the aerosol cloud was obtained compared to the pure drug particles.

5.2.3 Crystallisation

Crystallisation separates a solid substance from solution by means of changing the physical or chemical properties of the material concerned. A compound may crystallise from a solution when its solubility in that solvent is exceeded, i.e. the solution becomes supersaturated. Supersaturation can be achieved in many ways, some of which involve the evaporation of the solvent, cooling the solution or by producing additional solute as a result of chemical reaction or a change in the solvent system (by the addition of a poor solvent for the crystallising material). Crystallisation provides an important means of preparing inspirable particles and has advantages over both micronisation and spray-drying in that it may produce particles of high crystallinity and of predetermined size, size distribution and shape. Crystallisation may be carried out in either conventional solvents such as ethanol or in a supercritical fluid (Shekunov and York, 2000).

The supercritical fluid crystallisation has become one of the most studied techniques for the production of dry powders for aerosolisation in recent years (Phillips and Stella, 1993). If a gas is compressed to such a high pressure that it has the same density as its liquid and both appear as a single phase, then the fluid obtained is called supercritical (SC) fluid (SCF). A typical SCF is liquid nitrogen. Many drugs can dissolve in supercritical fluids, depending upon the physico-chemical properties of both the solute and the specific SCF. As with classical methods, crystallisation of a compound from a SCF solution can be carried out by either allowing the solvent to evaporate, or by adding a non-solvent or antisolvent (a solvent that is a poor solvent for the solute but is miscible with the SCF) to the solution to reduce the solubility of the solute, or by adding the solution to an antisolvent to force precipitation or crystallisation.

SCFs possess some unique qualities and have advantages over conventional solvents in producing fine particles which are particularly suitable for inhalation. Unlike conventional solvents, supercritical fluids have liquid-like densities with very large compressibilities, and viscosities intermediate between the gas and liquid extremes. The solvent power of a SCF can be manipulated by changing the temperature and/or pressure, thus allowing the system's solvent attributes to be adjusted or manipulated in a continuum. SCFs have been employed to prepare ultrafine particles of some pharmaceutical compounds, particularly steroids, which are polymorphic and their degree of crystallinity may be altered by mechanical micronisation (Otsuka and Kaneniwa, 1990). Their reasonable solubility in SC CO_2 makes them attractive candidates to be investigated for evaluating the potential of SCF crystallisation as a replacement for mechanical micronisation for preparing inspirable particles. Larson and King (1986) were able to precipitate a steroidal compound with approximately the same size distribution as the material obtained by mechanical micronisation. X-ray diffraction patterns indicated that drug particles retained crystalline structure after recrystallisation from a SCF but variations in the intensity of the patterns suggested some amorphous content. Mechanically micronised and SCF recrystallised phenacetin particles were found to possess similar melting points, thermograms and dissolution kinetics but showed obvious differences in physical appearance and specific surface areas (Loth and Hemgesberg, 1986). However, one of the major problems associated with the use of a SCF to produce inspirable particles is the likelihood of the fine particles forming larger aggregates, due to the high surface energy intrinsic to ultrafine particles (Phillips and Stella, 1993). This may not only lead to a reduction in the true particle size of the bulk powders but also to an increase in the interparticulate forces between the drug and/or the carrier particles.

Rapid recrystallisation may also lead to the production of metastable crystal forms, which are likely to transform to stable forms during further processing and storage. This possible transformation is undesirable for effective and reproducible drug delivery. Further work in this field is needed to investigate the crystalline properties and long-term stability of the physico-chemical characteristics especially the size, shape and deposition profiles both *in vitro* and *in vivo*, of SCF-recrystallised drug particles. Another problem in the use of a SCF may be that as a result of the low solubility of many drug entities in a supercritical liquid, the use of gas antisolvent (GAS) recrystallisation may be necessary. This technique requires the drug to be dissolved in an organic solvent and this is then precipitated by a SCF (Gallagher-Wetmore *et al.*, 1994). This procedure may greatly widen the application of supercritical fluids in the preparation of inspirable particles.

Crystallisation of drug particles from ordinary solvents under carefully controlled conditions may also provide a powerful alternative to mechanical micronisation, spray-drying and supercritical crystallisation in preparing inspirable fine particles. In order to obtain crystals of a suitable size and shape for inhalation, the basic crystallisation procedures may involve dissolving the drug in a solvent and then either adding the solution to an antisolvent, or adding the antisolvent to the drug solution to initiate crystallisation of the drug. The addition of the drug solution to an antisolvent may be preferential since by adding the former solution slowly to a vigorously stirred antisolvent, crystallisation can be so rapid that only small crystal nuclei of the drug are obtained. For example, salbutamol sulphate was crystallised by adding its aqueous solution to absolute ethanol to obtain elongated crystals (needle-like) with an MMD of 5.49 μm (Zeng, 1997). After blending with Lactochem® lactose, the recrystallised salbutamol sulphate gave a fine particle fraction (<6.4 μm) of 22.8%, which was more than double the FPF (10.8%) of micronised salbutamol sulphate with an MMD of 4.79 μm. The possibility of preparing elongated fine particles of drug by recrystallisation may offer another opportunity of optimising drug delivery from dry powder aerosols since elongated objects such as fibres and needle-like crystals have aerodynamic diameters largely dependent upon their shorter axes but practically independent of their lengths (see chapter 4). These particles may be expected to have smaller aerodynamic diameters as compared to more spherically shaped particles of a similar geometric size, since they depend largely on their shorter axes, which are much smaller than the mean diameters of spherical particles of a similar size. It can therefore be predicted that such axes will be more uniformly distributed than more isometric particles. Once inhaled, the elongated particles may possess a high selectivity of deposition in the respiratory tract. In a similar manner to fibre pollutants, elongated particles may deposit in the airways through interception (Timbrell, 1965). Moreover, recrystallised drug particles may have higher crystallinity, lower surface energy and electrostatic charge than micronised particles. Also, the shape of crystallised particles may be more easily controlled than micronised particles. The use of recrystallised needle-shaped particles may therefore provide another means of improving drug delivery from dry powder aerosols, although further work remains to be carried out to validate this hypothesis.

5.3 CARRIER PARTICLES

In most DPIs, inert coarse carrier particles are mixed with the fine drug particles such that the latter adhere to the former. The coarse carrier particles are used to aid the flow and dispersion properties of the fine drug particles. A number of different carriers have been used although lactose has been employed most frequently, solely because it has a history as a widely used and safe excipient in solid dosage forms (Timsina *et al.*, 1994). The carrier particles are designed to be of such a size that after inhalation, most of them remain in the inhaler or deposit in the mouth and upper airways. In order to reach the lower airways, drug particles must therefore dissociate from the carrier particles and become redispersed in the air flow. The redispersed drug particles may then undergo inertial impaction in the mouth, on the back of the throat and the upper airways. Some particles (2–5 μm) are likely to reach and deposit in the lower airways through gravitational sedimentation, interception and Brownian diffusion.

Particles used as carriers are mainly sugars and include lactose, sucrose and glucose. Studies are currently being conducted to evaluate the suitability of different sugars for use in DPIs at King's College London since the physico-chemical properties of the carrier determine, to a large extent, whether the formulation is successful or not. So far, lactose is the only acceptable carrier for dry powder aerosols in the USA, but some sugars obtained as crystalline sieve fractions also appear to be suitable and these include trehalose and mannitol (Byron *et al.*, 1996).

The material used as a carrier for inhalation aerosols should be readily available in an acceptable pharmaceutical grade, be chemically and physically stable and not interact with the drug substance. Most importantly, the carrier must not have any side-effects and once delivered to the respiratory tract, it should be easily metabolised or rapidly cleared from the airways. The carrier must not have any unpleasant taste or odour and it should be inexpensive, easy to prepare and available in a crystalline form with low hygroscopicity. Ideally, the carrier particles should ensure maximal dose uniformity and delivery efficiency of drug particles. Much work needs to be done to find a carrier that can meet all the above requirements. However, lactose, as stated above, has been most commonly used to aid the drug flow and dispersion (Byron *et al.*, 1990) because it fulfils most of the aforementioned criteria. It is possible to obtain lactose in a wide spectrum of particle size distributions, but a typical particle size of lactose when used as the carrier for inhalation aerosols is between 63 and 90 μm (Bell *et al.*, 1971).

Although the carrier particles are mostly prepared by crystallisation, spray-drying has also been employed to prepare lactose particles. As in the case of spray-drying of drugs, lactose produced in the same way might be spherical in form but mainly amorphous in nature (Kibbe, 2000). When used as the carrier for drug particles for inhalation, spray-dried lactose was found to produce a much lower respirable fraction of drug than when crystalline lactose was employed (Kassem, 1990). From the forgoing, it appears that this is largely attributable to the greater interparticulate forces between spray-dried lactose and drug particles.

5.4 CARRIER-FREE FORMULATIONS

Although carrier–drug mixtures represent the most common formulations, the use of aggregates composed solely of drug particles could avoid some of the problems encountered with the formulations where coarse carrier particles are involved. The Turbohaler® designed by Astra is a typical example of a carrier-free dry powder inhaler device where loosely agglomerated drug particles are employed as the powder formulation. Carrier-free formulations are especially advantageous over the carrier–drug mixtures for patients who are intolerant to carriers such as lactose. However, it is more difficult to process carrier-free formulations than the drug–carrier mixtures since the inhaled drug normally requires a unit dose of the order of micrograms. In addition, since this means that the amount of material to be dispensed is small, after inhalation a patient may not 'feel' that a dose has been delivered. An individual may therefore take more doses than required, leading to over-dosing.

Agglomeration of fine drug particles to produce a powder with a size around 100-μm is often employed to solve the flowability problems associated with fine drug particles (Bell *et al.*, 1971). Drug particles can be aggregated by conventional granulation or spheronisation techniques using a small amount of a suitable solvent (Ganderton, 1992). The essential requirement for such agglomerates in aerosols is that deagglomeration must occur readily upon inhalation. Therefore, the interaction between the composite particles within the agglomerates must be weak for proper delivery, yet strong enough to withstand processing. The type of solvent used, the ratio of solvent volume to powder weight and the size fraction of the agglomerates, all significantly affect deagglomeration, dispersion and delivery of drugs to the lung from such formulations (Kassem *et al.*, 1991). The method used to prepare the agglomerates is also critical in determining their size, size distribution, shape and density.

5.5 PARTICULATE INTERACTIONS BETWEEN THE DRUG AND CARRIER PARTICLES

As discussed previously (chapter 1), interparticulate forces between solid particles are mainly the result of van der Waals forces, electrostatic forces, capillary forces and mechanical interlocking (Rietema, 1991). Although van der Waals forces are dominant over other forces under normal conditions for most pharmaceutical ordered mixes (Staniforth, 1985), the contribution of electrostatic forces to the overall interparticulate forces will increase as the particle size decreases (Bailey, 1984). This is mainly because particle mass decreases at a greater rate than particle surface area, resulting in smaller particles having higher charge-to-mass ratios than larger particles. Interparticulate forces due to electrical charge will decrease with an increase in the relative humidity of the surrounding environment. Under high relative humidities (e.g. more than the critical relative humidities of interacting particles), water vapour will start to condense on the particle surface and thus increase overall electrical conductivity of the particles, leading to a rapid electrical charge dissipation. Above the critical relative humidity, the conductivity of the system is increased so rapidly that electrical forces may become negligible (Boland and Geldart, 1972). Although materials with different water affinities exhibit different critical ranges, which may vary from 25 to 70% relative humidity, for most pharmaceutical ordered mixtures electrostatic charge becomes insignificant when the relative humidity exceeds 65%. Higher relative humidities may thus appear to be favourable but if the powders are exposed to too high a relative humidity, this will result in water vapour condensation onto the particles. However, if it becomes excessive it will in turn increase the interparticulate interaction due to capillary forces (chapter 1), increase powder physico-chemical reactivity and give rise to the possibility of growth of microorganisms in the powder formulations. Therefore, during powder processing, the environmental relative humidity has to be carefully controlled such that any tribo-electrostatic charge can be sufficiently dissipated without bringing about significant water vapour condensation within the powder systems. However, exposure of the powdered material to extreme environmental humidities should be minimised throughout powder storage when electrical charging is less of a problem. In a typical aerosol powder formulation, the drug particles (2–5 μm) are adhered to a coarse carrier (63–90 μm), the entrapment or incorporation of drug particles within any existing surface cavities of which may also markedly increase the overall adhesion forces between the drug and the carrier.

For drug–carrier formulations, dissociation of the drug from the carrier particles and their subsequent dispersion are critically determined by the drag forces experienced in the inhaled air stream, the cohesive forces between drug particles and the adhesive forces between drug and carrier. Any means of increasing the drag forces generated by the inhaled stream, such as increasing inhalation flow rate or turbulence, will result in more drug particles being dissociated from the carrier particles and dispersed into the air stream, leading to higher delivery efficiency of the drug to the lower respiratory tract. However, most drugs delivered by inhalation are intended for the treatment of lung diseases and the maximum flow rates achievable by the patient may be limited due to their impaired lung function. Even after inhalation at extremely high flow rates, many drug powders are still found to remain adhered to carrier particles (Kassem, 1990). This portion of drug particles is likely to deposit in the mouth, back of the throat and upper airways and will eventually be swallowed. However, increasing the inhalation flow rate will increase the probability of dispersed drug particles being deposited in the upper airways. Therefore, it is not always possible to improve drug delivery efficiency to the lower airways by means of increasing the flow rates of the inspiratory air stream. An ideal strategy would involve the reduction of interparticulate forces in combination with an attempt to increase the externally applied forces with a view to ensuring maximum drug dissociation and dispersion on inhalation and consequently, achieving an optimised dose of drug delivered to the lower airways.

Powder flow is of importance in the formulation of dry powders for inhalation since it is a contributory factor in ensuring uniformity of fill weight and can determine packing in both single-shot (capsule) systems and multi-dose reservoir systems. Higher flowability will also lead to higher uniformity of powder packing. Apart from particle size and shape, interparticulate forces will also influence the flowability of powders (Staniforth, 1982a). In order to obtain a maximised flowability of powders, the interparticulate forces should be minimised.

5.5.1 Morphology of carrier particles

5.5.1.1 Particle shape

The morphology of carrier particles can play an important role in determining the particulate interaction between the drug and the carrier and thus it may influence powder properties such as flowability, mixing uniformity, powder dispersion and deaggregation. The majority of carriers for dry powder aerosols are crystalline sugars, the morphology of which may change with the crystallisation conditions. For example, α-lactose monohydrate has been observed to exist in a wide variety of shapes, depending on the conditions of crystallisation. Both the initial lactose concentration and crystallisation temperature as well as the type of water-miscible organic solvent used to induce crystallisation were shown to affect the morphology of the resultant lactose crystals (Zeng *et al.*, 2000a, 2000b). Lactose crystals of different shape were shown to produce different *in vitro* deposition profiles for salbutamol sulphate (Zeng *et al.*, 1998, 1999). However, there have been no strict criteria established to control the particle size and morphological properties of the carrier particles employed for inhalation aerosols. Variation in such factors affecting carrier particles may be one of the major causes of the batch-to-batch variation in drug delivery encountered for most dry powder aerosol formulations. Therefore, the morphology and particle size of the carrier particles should be carefully controlled so as to optimise the delivery efficiency and reduce batch-to-batch variation of dry powder aerosols.

5.5.1.2 Surface smoothness

Particle surface smoothness is an important factor in determining particulate interaction either via cohesion or adhesion. Most commercially available carrier particles have undergone processing such as mechanical milling. It is not unusual for such a treatment to introduce asperities onto the carrier particle surface that are of a large scale relative to the micronised drug particles. During mixing, some drug particles are likely to be entrapped within these surface asperities. The portion of drug particles entrapped into the carrier surface asperities is unlikely to become dislodged from the carrier particles under normal inhalation conditions since they may adhere to the carrier particles via mechanical interlocking. Such interaction is several orders of magnitude greater than can be achieved by other interparticulate forces, such as van der Waals forces. The more asperities a particle surface has, the more drug particles are likely to be entrapped and the less drug may be detached from the carrier. Thus, increasing the surface smoothness of carrier particles may decrease the portion of drug particles that are entrapped in the cavities of carrier surface and, consequently, reduce the overall interparticulate forces between the drug and carrier particles. Surface asperities may also act as a shelter for any entrapped drug, rendering it non-available to the drag forces generated by the inhaled air stream and thus further reducing drug detachment. For example, the respirable fractions of salbutamol sulphate, delivered from three kinds of lactose particles having different surface smoothness, have been compared (Kassem, 1990). Recrystallised lactose particles were prepared such that they possessed highly smooth surfaces. The surface roughness or rugosity was assessed as a ratio of the surface area derived by air permeability measurements to the theoretical surface assuming the particle was perfectly spherical. The fraction of the delivered dose that was less than 5 μm was reported (Ganderton, 1992) to be higher for formulations employing lactose particles with a lower value of rugosity of the carrier than in the case of lactose particles with a higher rugosity (Table 5.2). However, the use of rugosity ratio to measure surface roughness is limited since rugosity as defined above is a factor which is determined not only by the microscopic surface texture of the particles but also by macroscopic shape. Particles of different

TABLE 5.2

The effect of surface roughness (rugosity) of lactose carriers on the respirable fraction of salbutamol sulphate after inhalation through a Rotahaler® at two different flow rates (data from Ganderton, 1992)

Lactose carrier	Rugosity	% respirable fraction	
		60 l min^{-1}	150 l min^{-1}
Crystalline	2.3	4.1	17.0
Recrystallised	1.2	23.0	42.0
Spray-dried	2.6	5.0	23.0

shape will produce different rugosity values even though they may possess the same surface smoothness. For example, an elongated particle that has a smooth surface still has a relatively high rugosity value, which may be mistakenly interpreted as the particle having a rough surface. Thus, higher rugosity ratios are indicative of rougher surfaces only when the particles considered have the same shape. Hence, more accurate methods, such as those employing fractal geometry, need to be employed in order to assess the surface smoothness of the particles (see Section 7.3.3).

Thus, although the improved respirable fraction was attributed to increased surface smoothness of the recrystallised carrier particles (Table 5.2), other factors such as particle shape and processing history may also have differed between the commercially available Lactochem® lactose and the batch of recrystallised lactose. For example, Lactochem® lactose, as supplied, had previously undergone a milling process whilst the recrystallised lactose did not undergo any such treatment (Kassem, 1990). Therefore, as discussed previously (section 5.2.1), such processing may have induced amorphous regions onto the particle surface. Furthermore, the recrystallised lactose and the regular crystalline lactose may have had different particle shapes. However, none of these factors were taken into account in these studies and the data were not subjected to sound statistical analysis (Kassem, 1990; Ganderton, 1992). Therefore, the direct comparison of drug deposition from a formulation containing Lactochem® lactose with that from formulations employing recrystallised lactose might not reveal the true effects of surface smoothness of the carrier particles on the deposition of the drug. In order to investigate the effects of specific morphological characteristics of the carrier particles on drug deposition, the contribution of all the other properties of the carrier particles to the adhesion forces of drug particles, ideally have to be minimised, or kept at a similar level. This was achieved in a recent study by comparing drug deposition from different batches of lactose particles that had undergone similar preparative procedures (Zeng *et al.*, 1997). Lactose crystals were prepared from aqueous solutions so as to have similar shapes but different surface smoothness. These crystals were fractionated so that the particle size was between 63 and 90 μm and then mixed with micronised salbutamol sulphate. *In vitro* deposition of salbutamol sulphate was measured using a twin impinger after aerosolisation at $60\,l\,min^{-1}$ via either a Rotahaler® or a Cyclohaler®. The fine particle fraction (FPF) of the drug was calculated in terms of both the recovered dose (RD) and the emitted dose (ED).

From Figures 5.2 and 5.3, it can be seen that increasing the surface smoothness of lactose carrier particles, as denoted by the 'surface factor', generally resulted in an increase in the FPF of salbutamol sulphate after aerosolisation at $60\,l\,min^{-1}$ from either a Rotahaler® or a Cyclohaler®. Increasing surface smoothness would be expected to decrease mechanical entrapment of fine drug particles within any asperities on the surface of the carrier. However, when the mechanically entrapped drug particles are available for deaggregation, further improvement in the surface smoothness may have a very complicated effect on the overall interaction between drug and carrier particles. For example, increasing the surface

CHAPTER 5

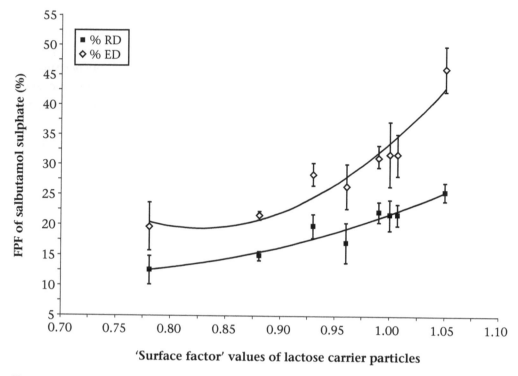

Figure 5.2: The relationship between the 'surface factor' of lactose particles and the fine particle fraction (FPF) of salbutamol sulphate aerosolised at 60 l min^{-1} via a Cyclohaler® expressed as a percentage of the recovered dose (RD) and emitted dose (ED) (error bars denote standard deviation, $n \geq 3$) (from Zeng, 1997).

smoothness may increase van der Waals forces and increase the sensitivity of the interactive systems to liquid bridging. Thus, increasing surface smoothness under these conditions may not result in a further increase in the deaggregation of fine powders.

5.5.1.3 Particle size

It is possible to produce lactose with a wide spectrum of particle size distributions but a typical particle size of lactose when used as a carrier for inhalation aerosols is between 63 and 90 µm (Bell *et al.*, 1971). Particles of different size ranges have different dispersion and deaggregation properties and, hence, the size of carrier particles will also affect the delivery of the drug. One study compared the effect of the size of lactose particles on the respirable fraction of drug–carrier ordered mixes (1 : 67.5) (Kassem *et al.*, 1989). Salbutamol sulphate and the Rotahaler® were used as model drug and inhaler device, respectively. The respirable fraction of the respective blends was measured *in vitro* using a cascade impactor at flow rates from 50 to 200 l min^{-1}. It was found that smaller carrier particles produced a higher respirable fraction of the drug at all flow rates. For example, increasing the carrier particles from <4.8 µm to a range of 63–90 µm reduced the poten-

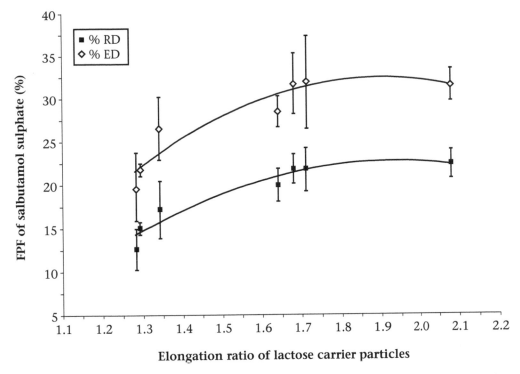

Figure 5.3: The relationship between the elongation ratio of lactose particles and the fine particle fraction (FPF) of salbutamol sulphate aerosolised at $60\,l\,min^{-1}$ via a Cyclohaler® expressed as a percentage of the recovered dose (RD) and emitted dose (ED) (error bars denote standard deviation, $n \geq 3$) (from Zeng 1997).

tially respirable, or fine particle, fraction of the drug from 17 to 4% after aerosolisation at $60\,l\,min^{-1}$. At $200\,l\,min^{-1}$, the fine particle fraction decreased from almost 50% for the smaller sized carrier particles to about 20% for the larger sized lactose. However, the influence of particle size of carrier particles on drug deposition may be drug-specific. In another study, a coarser lactose (53–105 μm) was shown to yield a higher respirable fraction of terbutaline sulphate than a finer lactose ($\leq$53 μm) after aerosolisation of the drug–carrier mixtures (1 : 49) also from a Rotahaler® (Byron *et al.*, 1990). It would also be expected that different inhaler devices may require different sized lactose particles in order to perform optimally. Since the particle size of the carrier is important in determining the powder flowability, dispersibility and delivery efficiency of drugs, it should be optimised for each device and powder formulation.

5.5.2 Electrical properties

In the case of powders for inhalation, the practical significance of electrostatic interactions is considered to be critical and more importantly, in the timely disruption of drug–carrier agglomerates, allowing reproducible delivery of respirable particles. In order to maintain a stable ordered mix, then a suitable magnitude of opposite charges of drug and carrier particles is desirable (Staniforth, 1994). However, excessive opposite charges may result in strong adhesive forces, leading to poor deaggregation of drug powders on aerosolisation, especially by children and patients with severe bronchoconstriction. Excessive forces may be overcome by the addition of a ternary material. For example, magnesium stearate was found to cause 'stripping' of drug particles from carrier particles in ordered mixes of salicylic acid and sucrose (Lai and Hersey, 1979). This phenomenon may be partly attributable to the saturation of active binding sites on sucrose by magnesium stearate and partly due to the change in the electrostatic properties of the constituent powders. Both salicylic acid and sucrose were found to develop electronegative charges when in contact with glass whereas magnesium stearate was strongly electropositive. Thus, magnesium stearate would be strongly attracted to the electronegative sucrose, competing for the active binding sites on the carrier particles and consequently, salicylic acid would be expected to become dislodged or relocated at weaker adherence sites on the sucrose surface.

The charge accumulated on the surface of a powder is dissipated over a certain period of time (Malave-Lopez and Peleg, 1985). The rate of charge decay is mainly dependent upon conductivity of the materials, contact with other conductive materials and the relative humidity of ambient air surrounding the particles. A material that has lower electron conductivity will need a longer period of time to dissipate its charge. Most pharmaceutical powders are insulators and hence charge usually decays at relatively slow rates. If powders are brought into contact with highly conductive materials, such as metals, charge will decay more quickly than when the same powders are in contact with insulators such as polymers.

The electrical charges on a lactose surface were reported to be dissipated within a few minutes (Staniforth and Rees, 1982). Thus, the triboelectrification which occurs during a preparation process might appear to be a minor factor affecting the adhesion of drug particles to carrier particles when the DPI is used by patients, since it usually takes several months for DPIs to be used in clinical practice after preparation. However, as mentioned previously, the strength of the van der Waals forces between macroscopic bodies depends critically on the contact area and separation distance between them. Thus initial electrostatic forces may be one of the major forces that bring the separate particles closer to a surface, producing an increased dispersion force which would connect the particle much more tightly to the surface even though the electrical charges may have long been dissipated.

The electrification of particles on aerosolisation is also an important concern. As stated previously, electrical charges resulting from turbulent air flow were shown to be more than 100 times stronger than their contact charges. The highly charged particles may coagulate with each other in an air stream or on impact with the surface of the inhaler due to electrical precipitation. This may be partly attributable to the loss of drug in the inhalers (>10% total dose). If bipolar (particles carrying opposite charge) charging occurs during aerosolisation, the charged, deaggregated drug particles may coagulate with each other to form large agglomerates in the air stream. Even individual charged particles may induce opposite image charges in the upper airways and accelerate the deposition of drug particles in this region. Powder characteristics as well as the design and composition of inhaler devices may influence the triboelectrification of drug particles during aerosolisation. For example, the relative electron-donor or electron-acceptor properties of drug and carrier and the composition of the inhalers may be decisive in the type of charge induced on the drug and carrier particles. Theoretically, it would appear to be advantageous for drug and carrier particles to bear the same charge on aerosolisation so that more complete detachment of the drug particles might be obtained.

The specific charges (i.e. the ratios of total charges to mass) of the powders, α-monohydrate lactose, micronised salbutamol sulphate and beclomethasone dipropionate particles, were found to increase linearly with the pressures employed to fluidise these particles in a steel or brass cyclone (Carter et al., 1982). Smaller lactose particles (45–90 μm) showed markedly higher specific charges than larger particles (355–500 μm) under similar fluidisation conditions. This latter phenomenon was attributed to the greater particle number density, in the case of smaller particles, providing an overall increase in available surface area for charge transfer. All the particles investigated showed different charging tendencies when they were in contact with either steel or brass (Figure 2.3). Stainless steel was found to impart a higher electronegative charge to lactose particles than brass, due to the lower work function of stainless steel. Salbutamol sulphate charged electropositively when in contact with brass but electronegatively when in contact with stainless steel. However, beclomethasone dipropionate particles showed the opposite charging tendency. Generally, the magnitude of electrostatic forces on these particles is much larger than that of gravitational forces but quite small as compared with dispersion forces. However, electrostatic forces become more significant as the particle size is decreased.

Due to the intrinsic requirement of generating particles with a size of less than 7 μm, drug particles for inhalation are highly cohesive and have poor flowability. This is usually overcome by blending the fine particles with coarse carrier particles such as lactose. Uniformity of mixing is a prerequisite for the uniformity of subsequent dosing and ideally, drug particles should be evenly distributed amongst the carrier particles after mixing. To achieve this, the adhesive forces between drug and carrier particles are required to be stronger than the cohesive forces between the fine drug particles. Thus, during powder mixing, higher adhesion forces are favoured in order to obtain the desirable uniformity of the ordered mix and increasing adhesion forces would lead to more even distribution of drug particles amongst the carrier particles. For example, the

CHAPTER 5

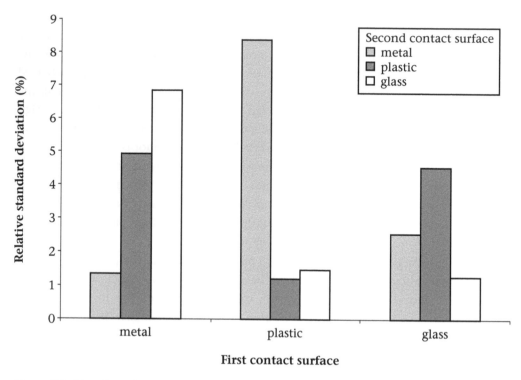

Figure 5.4: The effect of contact surfaces during powder blending on the homogeneity of salbutamol sulphate and lactose blends (modified from Staniforth, 1995).

adhesion forces can be increased by optimising the electrostatic forces between the drug and carrier particles by means of imparting opposite charges on these particles and this has been confirmed (Staniforth, 1995) in the case of a salbutamol sulphate and lactose blend. Lactose and salbutamol sulphate were blended in two stages: the first involved the mixing of lactose alone in either a glass, metal or plastic container, followed by the second stage of combining the drug with lactose in glass, metal or plastic containers. Homogeneity of the drug content in each powder mixture was then analysed and expressed as the relative standard deviation (RSD) as shown in Figure 5.4.

It can be seen that the mixture prepared by blending lactose in a plastic container then mixing with the drug in a plastic container produced the lowest RSD, indicating the highest homogeneity of drug content. On the other hand, the mixture prepared by first blending lactose in a plastic container followed by mixing with the drug in a metal container had the highest RSD, i.e. the lowest homogeneity of drug content in the powder formulation. These phenomena were attributed to the different charges of lactose and salbutamol sulphate after coming into contact with the different container materials. Salbutamol sulphate particles charge positively in all contact environments, whereas lactose particles charge negatively in contact with glass and metal but positively when in contact with plastic (Staniforth, 1994). Thus, during mixing in a metal container, lactose and salbutamol sulphate will adopt opposite electrical charges and this may result in stronger adhesion forces between the drug and carrier that will result in a poorer mixing efficiency. However, when mixing in a plastic container, both lactose and salbutamol sulphate charge electropositively and the resultant weaker adhesion between the drug and carrier may be the major reason for the better mixing efficiency.

Excessive charging should be avoided in order to minimise electrical precipitation of drug particles in the upper airways. In order to increase turbulence, resistance to air flow is normally introduced in the design of inhaler devices (Timsina *et al.*, 1994). This may lead to an increased triboelectrification of the deaggregated particles since the probability of the particles hitting or colliding with a surface is increased. The triboelectrification of insulators is a very complicated process and more work needs to be done with regard to the charging properties of drug and carrier particles. The coagulation and deposition profiles of charged particles should be the subject of further investigation.

The charge induced on the plastic inhaler device by the patient has not been extensively studied. Inter- and intra-subject variability in carrying charge is apparent as evidenced by the electrostatic discharges to earth that occur periodically by the individuals touching metal objects. It is interesting to speculate that a patient operating their inhaler in rubberised footware on a nylon carpet and in a dry atmosphere may be treated less effectively than under ambient conditions which ensure less charge retention by the individual. Reduced bioavailability might result from a higher proportion of drug being retained from the more highly charged device and adhered to the carrier. Indeed in at least one pharmaceutical company as part of the Standard Operating Procedure, technical staff earth themselves through the water pipes before determining emitted dose or fine particle fraction so as to reduce variability in results. However, to date, the instruction for asthmatic patients to chain themselves to the cold water pipes before operating their dry powder inhaler does not appear to be included routinely in patient information leaflets.

5.5.3   Drug–carrier formulations

5.5.3.1   Drug to carrier ratio

Solids with a high surface energy have a high tendency to adsorb other materials onto their surfaces forming strong bonds (Sutton, 1976). A material shows its highest surface energy when its surface is completely clean. Thus, any contamination of particle surfaces by gas, liquid or solid particles would reduce the adhesive forces between adhered particles as long as liquid and/or solid bridging is not involved. This may have significant implications for the theory of ordered mixes. Hersey (1975) suggested that some areas of carrier surfaces are devoid of binding sites whilst other areas possess strong binding sites. On clean surfaces, these strong binding sites are available for adhesion and would induce strong forces to adhered particles. However, on contaminated surfaces, these strong binding sites may be more likely to be saturated by the contaminants and thus more particles may be adhered to less strong binding sites and lower adhesive forces between the particles and carrier surface may be expected.

Although drug–carrier blends are usually more readily fluidised and deaggregated during inhalation than neat drug, increase in the carrier concentration should not exceed a certain limit beyond which there is, generally, a measurable decrease in the delivery efficiency of the drug. For example, Kassem (1990) compared the respirable properties of powder blends of salbutamol sulphate and lactose prepared in the ratios of 1 : 6.4, 1 : 33.5, 1 : 67.5 and 1 : 135. The highest ratio was calculated with a view to the drug particles forming a close-packed monolayer on the carrier surface. The respirable fraction from a Rotahaler®, measured with a cascade impactor, was found to increase significantly as the amount of drug increased and the effect was more pronounced at higher flow rates. These results were indicative of active binding sites on lactose crystals. The increased respirable fractions at higher drug concentrations was attributed to the relatively low percentages of drug powders adhered to these sites (Ganderton and Kassem, 1992). The optimal drug to carrier ratio is formulation and device dependent. Different particle sizes of drug and carrier, surface characteristics of these components such as shape and roughness, powder processing conditions, residual moisture level, etc. may require different optimal ratios of drug to carrier. Therefore, in order to obtain an optimised drug–carrier ratio, all these factors, as well as design of inhaler device, patient's inspiratory flow rate and therapeutic purpose of the drug, have to be considered.

5.5.3.2 Ternary materials

The use of an increased amount of drug particles to decrease the adhesion forces and hence increase the respirable fraction is not always practical, since the dose of drug administered is primarily dependent upon its therapeutic index. An alternative means of increasing the respirable fraction may be through the use of a ternary component, such as magnesium stearate, to saturate the active binding sites of the carrier particles before blending with drug powders. For example, the addition of a small amount of magnesium stearate to the ordered mixes of sucrose and salicyclic acid was found to reduce the homogeneity of the mixes and, as stated above, this was again attributed to the stripping action of magnesium stearate (Lai and Hersey, 1979). A similar principle has been employed to reduce the adhesion of salbutamol to lactose carriers and, consequently, a higher respirable fraction of the drug on aerosolisation of the formulation was obtained (Kassem, 1990). Another strategy proposed to improve drug detachment from carrier particles has been the incorporation of fine particles of L-leucine into the powder formulations (Staniforth, 1996) with a view to the fine particles occupying the sites of strong adhesion. However, the use of a third component in powder formulations for drug delivery to the lung will require stringent toxicological testing. Another strategy might be to include more carrier particles, smaller in size than the coarse carrier particles but larger than the drug particles. Such intermediate-sized carrier particles might be expected to prevent the direct interaction of the two components and, consequently, improve detachment of the drug from the carrier.

5.5.3.3 Fine particles of carrier

The effects of fine carrier particles on drug deposition from dry powder aerosols has been investigated extensively using lactose–salbutamol sulphate as a model mixture (Zeng et al., 1996). Thus, lactose particles (63–90 μm), treated initially with compressed air in order to remove some of the fine particles, were mixed with salbutamol sulphate (6.4 μm) in a ratio of 1 (drug): 67.5 (lactose), w/w. Drug deposition from the formulation was determined using a twin impinger and the results were compared with those obtained from the same batch of lactose before compressed air treatment.

The FPF of salbutamol sulphate, as measured by a twin impinger, was found to decrease significantly ($p < 0.05$) after the lactose particles had been cleaned with compressed air (Figure 5.5). Formulations based on untreated lactose produced an FPF of

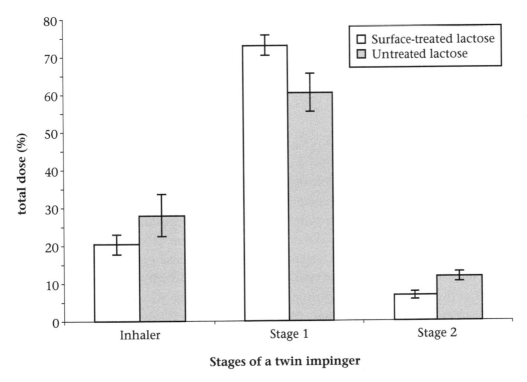

Figure 5.5: Deposition of salbutamol sulphate in a twin impinger after aerosolisation at 60 l min^{-1} via a Rotahaler® (from Zeng et al., 1996).

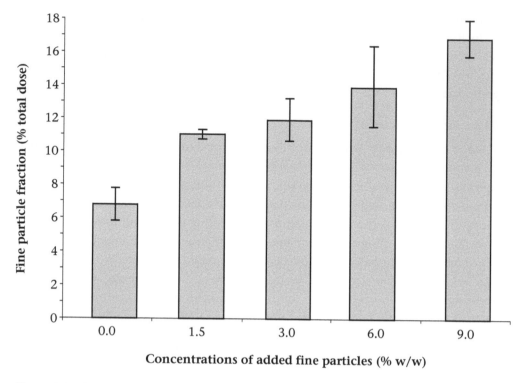

Figure 5.6: Effects of amount of added lactose particles (MMD 4.96 μm) on fine particle fraction of salbutamol sulphate after aerosolisation at 60 l min⁻¹ via a Rotahaler® (Zeng, *et al.*, 1996).

11.8 ± 1.1%, which was over 70% greater than the value of the FPF (6.8 ± 1.1%) with treated lactose. The fraction of the drug collected in stage 1 was on the other hand significantly higher ($p < 0.05$) with treated lactose (72.8 ± 2.8%) than with untreated lactose (60.2 ± 5.3%). The use of compressed air was shown to produce a cleaner lactose particle surface (Tee, 1996) when compared to that before treatment. Stripping of fine lactose particles from the coarse carrier with compressed air produced more adhesion sites for the drug particles. This would be expected to increase the overall particulate interaction between the drug and coarse carrier particles, which would in turn reduce the FPF of the drug. Fine lactose particles (MMD ~5 μm) were added back to the cleaned coarse carrier and the more that were added, the higher the FPF of the drug (Figure 5.6). The improved FPF of salbutamol sulphate with increased concentration of added fine particles would be expected to saturate more active binding sites, fill more surface crevices and form multiple layers of fine particles on the coarser particle surface. All these mechanisms would reduce the interaction between the drug and carrier particles.

5.5.3.4 Ternary ordered mixes

In conventional binary ordered mixes, particulate interaction between the drug and carrier may be so strong that most DPIs may not generate sufficient air drag force to dislodge the drug from the carrier particles when drug particles are directly adhered to the coarse carrier particles. Addition of a small amount of intermediate-sized lactose particles to powder formulations containing lactose and salbutamol sulphate was found to increase significantly the fine particle fraction (FPF) of the drug (Zeng et al., 1996, 1998). Thus, the use of ternary ordered mixes (TOM) composed of micronised drug, intermediate-sized and coarse carrier may provide a better means of drug delivery from dry powder aerosols than the binary ordered mix.

For example, ternary ordered mixes (TOM) of coarse lactose (88.9 μm), intermediate-sized lactose (8.1 μm) and salbutamol sulphate (5.8 μm) were prepared in a ratio of 64.1 : 3.4 : 1 using different sequences of powder mixing as shown in Table 5.3. The *in vitro* deposition of lactose and salbutamol sulphate was determined using an Andersen cascade impactor (ACI) after aerosolisation at a flow rate of $60 \, l \, min^{-1}$ from a Rotahaler® (Zeng et al., 1999). TOM, regardless of the mixing sequences, produced a fine particle fraction (FPF) of salbutamol sulphate, which was significantly higher ($p < 0.01$, ANOVA) than that from a binary ordered mix (BOM) composed of coarse lactose and salbutamol sulphate (67.5 : 1) only (Figure 5.7). Different mixing sequences of the individual components of TOM resulted in different FPFs of both the drug and the carrier. For example, powder formulations prepared by first blending the intermediate-sized lactose with coarse lactose then mixing with the drug (TOM1) produced a higher FPF of salbutamol sulphate than formulations prepared using other orders of mixing the various components (TOM2 and 3). Interestingly, TOM1 produced the lowest FPF of lactose whilst TOM3, prepared by first blending salbutamol sulphate with coarse lactose and then mixing with the intermediate-sized lactose, produced the

TABLE 5.3

Mixing sequences employed to prepare the ternary ordered mixtures composed of salbutamol sulphate, coarse and fine lactose

Formulations	Mixing sequences	
	Initial blend[a]	Final component[b]
TOM1	Coarse lactose + fine lactose	Salbutamol sulphate[c]
TOM2	Salbutamol sulphate + fine lactose	Coarse lactose[c]
TOM3	Salbutamol sulphate + coarse lactose	Fine lactose[c]
BOM	Salbutamol sulphate + coarse lactose	

[a]Components mixed for 30 min.
[b]Final component mixed with initial blend for further 30 min.
[c]Blending components.

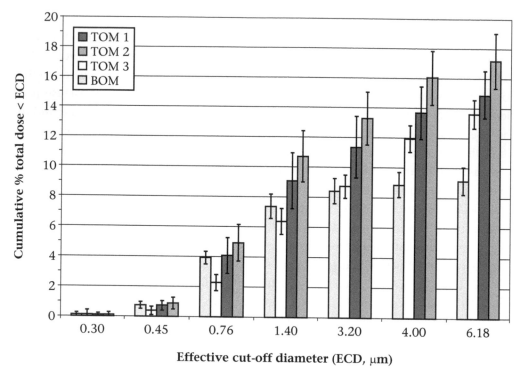

Figure 5.7: Particle size distribution of salbutamol sulphate after aerosolisation at 60 l min⁻¹ via a Rotahaler® (after Abeer *et al.*, 1997) (see text for abbreviations).

highest FPF of lactose (Figure 5.8). Therefore, more salbutamol sulphate might be expected to adhere directly to the coarse carrier in TOM3 than in TOM1 whilst any added intermediate-sized lactose might follow the reverse trend. Consequently, TOM1 produced the higher FPF of salbutamol sulphate but the lower FPF of lactose in comparison to TOM3. TOM2 was prepared by first blending salbutamol sulphate with the intermediate-sized lactose and then mixing with the coarse lactose. Thus, it produced an intermediate FPF of lactose and salbutamol sulphate.

A ternary ordered mix composed of salbutamol sulphate, intermediate-sized and coarse lactose may be superior to the binary ordered mix containing the drug and coarse carrier only, for a more efficient delivery of drug from dry powder aerosols. The mixing sequence of the various components may also affect the delivery of both the drug and carrier. By controlling the drug formulation and mixing processes, it may be possible to improve drug delivery efficiency without introducing too many fine carrier particles into the lung.

The FPF of beclomethasone dipropionate (BDP), a hydrophobic drug, was also shown to be dependent upon the processing conditions when mixing time was short but as the time was lengthened this dependency diminished (Zeng *et al.*, 2000c).

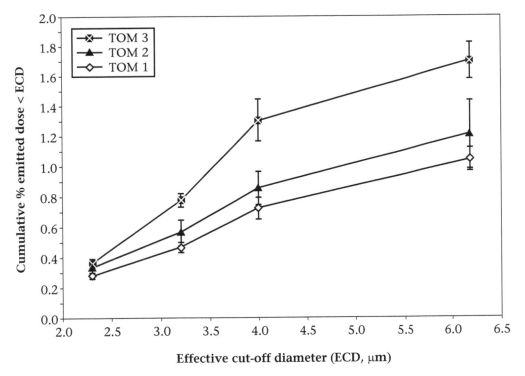

Figure 5.8: Particle size distribution of lactose from different formulations after aerosolisation at 60 l min^{-1} from a Rotahaler® (after Abeer *et al.*, 1997) (for abbreviations see text).

5.5.4. Other physico-chemical properties

5.5.4.1 Deformability

The deformability of a material is largely dependent upon its crystalline state since crystals are known to be much harder than amorphous forms under similar temperatures. Thus, a crystalline drug can be expected to have much lower adhesion to a surface than its amorphous forms. For example, spray-dried salbutamol sulphate was found to be softer and more 'sticky' than the micronised drug which is more crystalline (Chawla *et al.*, 1994). After mixing with lactose carrier particles, the spray-dried drug particles would be expected to exhibit higher adhesive forces to the carrier than micronised salbutamol sulphate. The spray-dried drug was shown, in turn, to have a lower respirable fraction after inhalation than the micronised drug even though the former had a more spherical form with a lower mass median diameter than the latter (Venthoye *et al.*, 1995). The decreased respirable fraction with spray-dried salbutamol sulphate is therefore likely to be largely due to the increased adhesive forces between the drug and the carrier particles. These increased adhesive forces are often produced by more extensive mixing process and this may also be partly attributable to an increased contact area between drug and carrier particles.

5.5.4.2 Chemical properties

As well as the physical properties of the interactive particles, chemical structure may also affect the interparticulate forces. Generally, materials having similar polarity or hydrophilicity adhere to each other more strongly than materials of different polarities (Podczeck *et al.*, 1996). Carriers produced from different chemical entities may exhibit different interparticulate forces with a specific drug particle. Thus, carrier particles made with different materials may produce altered deposition profiles with the same drug particles. For example, the FPF of salbutamol base was shown to be the highest after aerosolisation from sorbitol, followed by crystalline lactose and maltose with the least FPF resulting from dextrose and spray-dried lactose (MacRitchie *et al.*, 1995). Differences in drug delivery could not be explained on the basis of chemical structure alone since each carrier could have possessed a different particle size and morphology. Nevertheless, different sugars have different polarities and hydrophilicities and therefore, would be expected to exert different adhesion forces to the drug particles, leading to different drug delivery profiles from formulations using these carriers. Further, the controlled crystallisation of different sugars may produce a variety of crystal habits and forms and this may provide a wealth of opportunities for optimising drug delivery from dry powder aerosols. Obviously, the various permutations of employing two different carrier types with different size distributions, both prepared under conditions of controlled crystallisation, offer extensive scope for future investigation.

5.5.5 Preparative process

The dependence of adhesion forces on the processing procedures and more specifically on mixing, is obvious. Most pharmaceutical drug particles are irregular in shape, and during extensive blending, particles may seek the orientation of smallest potential energy. This arrangement involves the shortest separation and largest contact area, especially for elongated and plate-like particles and thus, relatively high adhesive forces can be expected to prevail under these conditions. Kulvanich and Stewart (1988) showed that the adhesive tendency of drug particles in a model drug–carrier system increased with blending time. These effects may be partly due to the fact that under extensive blending, drug particles would progressively move to positions of greatest stability (Ganderton and Kassem, 1992), although an increased triboelectrification of the particles may also contribute to the increased adhesion. Therefore, the mixing procedures may also need to be clearly defined so as to ensure uniformity of mixing without substantially increasing the interparticulate forces between the drug and carrier particles.

The effect of applied pressure and 'ageing' on adhesion force may also have significant importance in the formulation of drug powders for inhalation. For example, during powder mixing processes, increasing the mixing time and/or force (e.g. the speed of rotation of the mixer) will blend drug and carrier particles more vigorously, which may be compared to an increase in applied pressure, and consequently might lead to increased adhesion of drug particles to the carrier. Work carried out at King's College London (KCL) has shown that extending mixing time and/or increasing mixing speed generally resulted in a reduction in the fine particle mass of salbutamol sulphate particles from lactose carriers. This phenomenon may be partly due to the increased force of adhesion generated between drug particles and the carrier under more intensive mixing, although other factors such as possible increases in triboelectrification may also partly contribute to the increased force. Although there are no reported studies examining the effect of 'ageing' of ordered mixes on the drug respirable fraction, some of our unpublished data, indicates that after storage in a desiccator for up to 4 months, the fine particle mass of salbutamol base from different carriers such as lactose decreased significantly in contrast to that produced with the freshly prepared powders. This reduction in the respirable mass may result from the increased adhesion force between drug and carrier particles after the prolonged period of contact. If such an increase in adhesion occurs, the drug particles would be less likely to detach from the carrier under similar inhalation conditions, resulting in a reduction in the respirable fraction.

5.5.6 Relative humidity

The influence of relative humidity on the performance of dry powders for inhalation has been a matter of concern for many years (Kontny et al., 1994) since many drugs and carriers employed are hygroscopic and exposure of these mixtures to high relative humidity results in an increase in adhesive forces due to capillary interaction. Therefore, humidity generally has detrimental effects on the performance of dry powder aerosols. If the formulation contains moisture-sensitive drugs and/or excipients, then exposure to high humidity conditions would complicate the processing conditions. Excessive humidity may also cause significant diffusion of moisture into packaging and device systems, leading to physico-chemical changes and microbial growth, either of which will cause serious efficacy and toxicity problems. Further, moisture uptake will greatly increase particulate interaction due to capillary forces. This will reduce the dispersion and deaggregation properties of the dry powder and overly moisture-sensitive compounds may be rendered unrespirable during transit of the powder particles in the respiratory tract due to high humidity in this region. For example, a powder formulation of micronised sodium salts of xanthine carboxylic acid, after equilibration at high humidity, gave virtually no respirable material in a fluidisation test. However, the acid was

CHAPTER 5

much less influenced by increased humidities in contrast to its salt (Chowhan and Amaro, 1977).

The importance of humidity control in handling and storage of dry powder aerosols cannot be overstated. Dry powders may have different sensitivities to the environmental humidity, depending on the hygroscopicity of the composite materials (drug and/or excipient). Generally, the salt form of a drug is more sensitive to moisture than the free-base form (Jashnani et al., 1995). Therefore, the functional performance of dry powder aerosols is also dependent upon the salt selection and product storage condition and it is preferable to choose a drug form and excipients which exhibit low sensitivity to environmental moisture.

Apart from the physico-chemical properties of dry powders, other factors to be considered when designing formulations for DPIs might include the use of gelatin capsule shells, multi- or unit dose packaging and the design of the inhaler device. Such factors may prove to be important in determining the sensitivity of a dry powder system to environmental moisture. Dry powders for inhalation are filled either directly into the packaging system as a unit dose (e.g. Diskhaler®) or multiple doses (e.g. Turbohaler®) or encapsulated in gelatin capsules prior to presentation as unit-dose (e.g. Inhalets®) or multi-dose (e.g. Rotacaps®) packaging. Since gelatin has a much higher moisture sorption capacity than many drug powders, encapsulation in gelatin would be expected to reduce the rate of increase in relative humidity inside the capsule shells and thus provide a protective barrier to external water vapour condensing on the encapsulated drug powders. For a multi-dose system, increasing the potential for moisture sorption inside the device by increasing the number of doses or desiccant would reduce the rate of increase in the relative humidity inside the packaging chamber (Kontny et al., 1994). On the other hand, unencapsulated, unit-dose drug powders (e.g. Diskhaler®) have the lowest moisture sorption capacity due to the lack of protection of gelatin shells, and the relative humidity in the dose chamber would rapidly achieve an equilibrium with the surrounding environment. Thus, encapsulated, multi-dose drug powders (Rotacaps®) are generally more stable with regard to the influence of environmental humidities than unencapsulated, unit-dose drug powders, such as those employed in the Diskhaler®.

Ordered mixes should maintain their uniformity during powder handling and storage. Segregation of drug particles may arise as a result of any vibration during powder handling such as filling, transport and storage. This may be due to the cohesive nature of the original fine powders, the large differences in particle size and the relative amounts of the formulation components and other mechanical properties of the drug and carrier particle (Staniforth, 1982b). Increasing the interparticulate forces between drug and carrier particles should improve the stability of the drug content in the ordered mixes (Hersey, 1979). The stability of ordered mixes toward mechanical vibration was shown to improve by either increasing the van der Waals forces (Staniforth and Rees, 1983), or electrostatic forces (Staniforth and Rees, 1982) or capillary forces

(Staniforth, 1985). High adhesion forces between drug and carrier particles are advantageous for tabletting and capsule filling, but this is not the case for dry powder formulations for inhalation. Increasing the adhesion forces will obviously decrease the extent of detachment and dispersibility of drug particles. Insufficient detachment and deaggregation of drug particles upon inhalation is often a major reason for the ineffective delivery of drugs to the lungs. These opposing factors have to be carefully balanced so as to ensure dose uniformity and delivery efficiency of drugs from dry powder aerosols.

5.5.7. Drug properties

The particulate interaction within either a blend of drug and carrier or a powder of drug particles alone is a function of the physico-chemical properties of the drug particles. As mentioned previously, spray-dried drugs usually produced smaller respirable fractions than micronised drugs of a similar geometric diameter although the former possessed a more regular, spherical shape (Chawla et al., 1994). This may largely be attributable to the deformable nature of spray-dried particles due to a relatively high amorphous content and to their hollow structure, in comparison to more crystalline, micronised particles. However, other properties such as density, porosity and hydrophilicity of drug particles may also have an important impact on dispersion and subsequently, deposition profiles of the drug in the airways.

5.5.7.1. Density and porosity

It has been mentioned before that in order to reach the lower airways, drug particles must have an aerodynamic diameter (d_a) of 2–5 μm. The d_a of a spherical particle relates to its geometric diameter (d) and density (ρ) by the following formula:

$$d_a = d\sqrt{\rho} \tag{5.1}$$

Therefore, d_a can be changed by altering d or ρ or a combination of both.

All the particle size reduction techniques mentioned so far are able to reduce the geometric diameter of drug particles without substantially changing particle density. Most therapeutic aerosols thus prepared have a density similar to the starting materials, roughly in the range of 0.5–1.5 g cm^{-3}. Thus, in order to obtain particles with $d_a < 5$ μm, the drug has to be micronised so that $d < 6$ μm. After blending with coarse carrier particles (e.g. 63–90 μm), the fine drug particles are likely to be entrapped in any valleys or crevices on the carrier surface (Figure 5.9). The entrapped particles would require a higher drag force from the air stream in order to be detached from the adherence sites. Even the particles that are not 'entrapped' into any surface crevices or valleys, may 'hide' behind any 'hills' or protruding parts of the carrier surface, which may protect these particles from being swept away by the air stream (Figure 5.9).

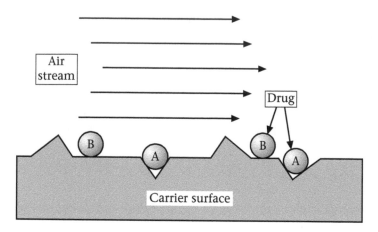

Figure 5.9: A diagrammatic presentation of some typical adhesions of drug particles onto a carrier surface (A represents particles that are entrapped in surface crevices whilst B denotes particles that 'hide' behind the protruding parts of the surface).

All these problems may be solved by preparing particles of relatively large geometric particle size. One approach is to prepare particles that have a large geometric diameter with an aerodynamic diameter sufficiently small to render the particles capable of penetrating deep into the lung. This can be achieved by employing the simple relationship between aerodynamic diameter, geometric diameter and particle density as outlined in equation (5.1). If a particle has a large geometric particle size (e.g. 10 μm), it can be made capable of penetrating deep into the lung by simply reducing the particle density, i.e. increasing the particle porosity. In other words, respirable particles may be prepared by reducing the particle density without drastically decreasing the particle size. Edwards *et al.* (1997) used a double-emulsion evaporation technique to prepare highly porous poly-lactic acid-co-glycolic acid (PLGA) particles with a mass density $<0.4 \text{ g cm}^{-3}$ and mean geometric diameter $>5 \text{ μm}$. After aerosolisation at 28.3 l min^{-1} via a Spinhaler® into a cascade impactor, the 'respirable' fraction (the percentage of the total particle mass exiting the DPI, recovered from the terminal stage of the impactor) was found to be $20.5 \pm 3.5\%$ and $50 \pm 10\%$ for non-porous particles with d of 3.5 μm and ρ of 0.8 g cm^{-3} and porous particles with d of 8.5 μm and ρ of 0.1 g cm^{-3}, respectively. Although the calculated values of d_a were similar, the porous particles were more readily dispersible than the non-porous particles. Since large porous particles are packed much less densely than the non-porous more compacted particles, the interparticulate forces within a powder composed of more porous particles can be expected to be much lower than those within a powder composed of the less porous particles, providing the former do not undergo significant deformation. Upon exposure to an air stream, the large porous particles would be expected to be subjected to a higher air drag force than the solid particles, due to the relatively large cross-sectional area of the former, leading to their improved dispersion and deaggregation after aerosolisation.

However, porous particles are likely to be more deformable than non-porous particles and this may occur during powder processing, particularly when they are mixed with a coarse carrier. If deformation does occur, then the interparticulate forces within a powder composed of porous particles may exceed those of the powder containing non-porous particles, since deformed porous particles will produce more contact areas and more mechanical interlocking. Therefore, caution must be taken to ensure that the morphology of porous particles does not change significantly during powder processing. Due to a higher specific surface area, porous particles would also be expected to be more susceptible to environmental humidity than less porous particles. Any water vapour uptake by these particles is detrimental to dispersion and deposition of the final product since this will increase interparticulate forces and particle density, leading to a decrease in dispersibility and an increase in the aerodynamic diameter. Upon entering the respiratory tract, porous particles will take up water vapour at a higher rate than solid particles of the same material. Thus, the density of porous particles will increase rapidly on travelling through the airways and this will consequently lead to an increase in the aerodynamic diameter of the aerosolised particles. Such an increase in density might lead to fewer particles reaching the alveolar region than originally predicted on the basis of *in vitro* data.

5.5.7.2. Surface modification

Apart from the use of porous particles, another novel approach has also been introduced to cope with the cohesiveness of micronised drug particles and this involves modification of the surface of drug particles (Kawashima *et al.*, 1998). Powders of highly hydrophobic drugs, such as beclomethasone dipropionate, are usually more cohesive than those of more hydrophilic drugs such as salbutamol sulphate if the component particles have a similar particle size and morphological features. For highly hydrophobic materials, any sorbed moisture may be condensed exclusively on the particle surface, leading to a maximum increase in capillary forces with increasing relative humidity (Fukuoka *et al.*, 1983). Thus, hydrophobic particles are more cohesive than hydrophilic particles under lower relative humidities since intact liquid bridges are more likely to form on the surface of hydrophobic particles than on those of hydrophilic particles. The cohesiveness of hydrophobic drug particles may therefore be reduced by coating the particle surface with hydrophilic materials.

Kawashima *et al.* (1998) modified the surface of an hydrophobic drug, pranlukast hydrate, with ultrafine nanospheres of an hydrophilic excipient, hydroxypropylmethylcellulose phthalate (HPMCP) by introducing drug particles to an aqueous colloidal HPMCP dispersion followed by freeze- or spray-drying of the resultant aqueous dispersion. The surface-modified drug was then blended with lactose and the mixture was aerosolised into a twin impinger at $60 \, l \, min^{-1}$ via a Spinhaler®. A two-fold increase in drug emission and a three-fold increase in the fine particle fraction were achieved from the powder composed of surface-modified drug as compared with the original unmodified material. Such an approach to improve the dispersibility of a hydrophobic drug has opened a new field in the development of dry powder formulations for inhalation and it would provide an important means of improving the delivery of many such drugs to the lower airways.

REFERENCES

ABEER, A.G., ZENG, X.M., MARTIN, G.P. and MARRIOTT, C. (1997) Effects of powder formulation and preparative processes on the in vitro delivery of salbutamol sulphate and lactose. In: *Proceedings of Drug Delivery to the Lungs VIII (London)*. December, 163–166.

AHLNECK, C. and ZOGRAFI, G. (1990) The molecular basis of moisture effects on the physical and chemical stability of drugs in the solid state. *Int. J. Pharm.* **62**, 87–95.

BAILEY, A.G. (1984) Electrostatic phenomena during powder handling. *Powder Technol.* **37**, 71–85.

BELL, J.H., HARTLEY, P.S. and COX, J.S.G. (1971) Dry powder aerosols. I. A new powder inhalation device. *J. Pharm. Sci.* **60**, 1559–1564.

BOLAND, D. and GELDART D. (1972) Electrostatic charging in gas fluidised beds. *Powder Technol.* **5**, 289–297.

BRIGGNER, L.E. (1994) Characterisation of powders by calorimetry. In: *Proceedings of the Second International Conference of Pharmaceutical Science and Technology, East Brunswick, NJ*. July 25–28.

BUCKTON, G., CHOULARTON, A., BEEZER, A. and CHATHAM, S. (1988) The effect of the comminution technique on the surface energy of a powder. *Int. J. Pharm.* **47**, 121–128.

BYRON, P.R., JASHNANI, R. and GERMAIN, S. (1990) Efficiency of aerosolization from dry powder blends of terbutaline sulfate and lactose NF with different particle size distribution. *Pharmaceut. Res.* **7**, S81.

BYRON, P.R., NAINI, V. and PHILIPS, E.M. (1996) Drug carrier selection–important physicochemical characteristics. In: *Proc. Respiratory Drug Delivery V*.

CARTER, P.A., ROWLEY, G., FLETCHER, E.J. and HILL, E.A. (1982) An experimental investigation of triboelectrification in cohesive and non-cohesive pharmaceutical powders. *Drug Dev. Ind. Pharm.* **18**, 1505–1526.

CHAN, H-K., CLARK, A., GONDA, I., MUMENTHALER, M. and HSU, C. (1997) Spray dried powders and powder blends of recombinant human deoxyribonuclease (rhDNase) for aerosol delivery. *Pharmaceut. Res.* **14**, 431–437.

CHAWLA, A., TAYLOR, K.M.G., NEWTON, J.M. and JOHNSON, M.C.R. (1994) Production of spray dried salbutamol sulphate for use in dry powder aerosol formulation. *Int. J. Pharm.* **108**, 233–240.

CHOWHAN, Z.T. and AMARO, A.A. (1977) Powder inhalation aerosol studies I: Selection of a suitable drug entity for bronchial delivery of new drugs. *J. Pharm. Sci.* **66**, 1254–1258.

EDWARDS, D.A., HANES, J., CAPONETTI, G., HRKACH, J., BEN-JEBRIA, A., ESKEW, M.L., MINTZES, J., DEAVER, D., LOTAN, N. and LANGER, R. (1997) Large porous particles for pulmonary drug delivery. *Science* **276**, 1868–1871.

FLORENCE, A. and SALOLE, E. (1976) Changes in crystallinity and solubility on comminution of digoxin and observations on spironolactone and oestradiol. *J. Pharm. Pharmacol.* **28**, 637–642.

FUKUOKA, E., KIMURA, S., YAMAZAKI, M. and TANAKA, T. (1983) Cohesion of particulate solids. VI. Improvement of apparatus and application to measurement of cohesiveness at various levels of humidity. *Chem. Pharm. Bull.* **31**: 221–229.

GALLAGHER-WETMORE, P., COFFEY, M.P. and KRUKONIS, V. (1994) Application of supercritical fluids in recrystallization: Nucleation and gas anti-solvent (GAS) techniques. In: BYRON P.R., DALBY, R.N. and FARR, S.J. (eds), *Respiratory Drug Delivery IV, Program and Proceedings, Virginia, USA.* 303–311.

GANDERTON, D. (1992) The generation of respirable cloud from coarse powder aggregates. *J. Biopharm. Sci.* **3**, 101–105.

GANDERTON, D. and KASSEM, N.M. (1992) Dry powder inhalers. In: GANDERTON, D. and JONES, T. (eds), *Advances in Pharamaceutical Sciences*, Vol. 6. London, Academic Press, 165–191.

GONDA, I. (1990) Aerosols for delivery of therapeutic and diagnostic agents to the respiratory tract. *Crit. Rev. Ther. Drug* **6**, 273–313.

GONDA, I. (1992) In: CROMMELIN D.J.A. and MIDHA, K.K. (eds), *Topics in Pharmaceutical Sciences.* Stuttgart, Medpharm Scientific, 95–115.

GUPTA, P.K. and HICKEY, A.J. (1991) Contemporary approaches in aerosolized drug delivery to the lung. *J. Control. Release* **17**, 127–147.

HERSEY, J.A. (1975) Ordered mixing: A new concept in powder mixing practice. *Powder Technol.* **11**, 41–44.

HERSEY, J.A. (1979) The development and applicability of powder mixing theory. *Int. J. Pharm. Tech. Prod. Mfr.* **1**, 6–13.

HICKEY, A.J. (1992) Summary of common approaches to pharmaceutical aerosol administration. In: HICKEY, A.J. (ed.), *Pharmaceutical Inhalation Aerosol Technology*. New York, Marcel Dekker, 255–259.

HICKEY, A.J., CONCESSIO, N.M., VAN, M.M. and PLATZ, A.M. (1994) Factors influencing the dispersion of dry powders as aerosols. *Pharm. Technol.* **18**, 58–82.

HINDS, W.C. (1982) In: *Aerosol Technology*. New York, John Wiley & Sons.

JASHNANI, R.N., BYRON, P.R. and DALBY, R.N. (1995) Testing of dry powder aerosol formulations in different environmental conditions. *Int. J. Pharm.* **113**, 123–130.

KASSEM, N.M. (1990) Generation of deeply inspirable dry powders. PhD thesis, University of London.

KASSEM, N.M., HO, K.K.L. and GANDERTON, D. (1989) The effects of air flow and carrier size on the characteristics of an inspirable cloud. *J. Pharm. Pharmacol.* **41** (Suppl.) p. 14.

KASSEM, N.M., SHAMAT, M.A. and DUVAL, C. (1991) The inspirable properties of a carrier-free dry powder aerosol formulation. *J. Pharm. Pharmacol.* **43** (Suppl.) p. 75.

KAWASHIMA, Y., SERIGANO, T., HINO, T., YAMAMOTO, H. and TAKEUCHI, H. (1998) A new powder design method to improve inhalation efficiency of pranlukast hydrate dry powder aerosols by surface modification with hydroxypropylmethylcellulose phthalate nanospheres. *Pharmaceut. Res.* **15**, 1748–1752.

KIBBE, A.H. (2000) *Handbook of Pharmaceutical Excipients*, 3rd edn. Washington, American Pharmaceutical Association and Pharmaceutical Press.

KONNO, T. (1990) Physical and chemical changes of medicinals in mixtures with adsorbents in the solid state. IV. Study on reduced-pressure mixing for practical use of amorphous mixtures of flufenamic acid. *Chem. Pharm. Bull.* **38**, 2003–2007.

KONTNY, M.J., CONNERS, J.J. and GRAHAM, E.T. (1994) Moisture distribution and packaging of dry powder systems. In: BYRON, P.R., DALBY, R.N. and FARR, S.J. (eds), *Respiratory Drug Delivery IV, Program and Proceedings, Virginia, USA*. 125–126.

KULVANICH, P. and STEWART, P.J. (1988) Influence of relative humidity on the adhesive properties of a model interactive system. *J. Pharm. Pharmacol.* **40**, 453–458.

LAI, F.K. and HERSEY, J.A. (1979) A cautionary note on the use of ordered powdered mixtures in pharmaceutical dosage forms. *J. Pharm. Pharmacol.* **31**, 800.

LARSON, K.A. and KING, M.L. (1986) Evaluation of supercritical fluid extraction in the pharmaceutical industry. *Biotechnol. Progr.* **2**, 73–82.

LOTH, H. and HEMGESBERG, E. (1986) Properties and dissolution of drugs micronised by crystallization from supercritical gases. *Int. J. Pharm.* **32**, 265–267.

MACRITCHIE, H.B., MARTIN, G.P., MARRIOTT, C. and MURPHY, L. (1995) Determination of the inspirable properties of dry powder aerosols using laser light scattering. *J. Pharm. Pharmacol.* **47**, 1072.

MALAVE-LOPEZ, J. and PELEG, M. (1985) Linearization of the electrostatic charging and charge decay curves of powders. *Powder Technol.* **42**, 217–223.

MUMENTHALER M., HSU, C. and PEARLMAN, R. (1991) Preparation of protein pharmaceuticals using a spray drying technique. *Pharmaceut. Res.* **8**, S59.

MUMENTHALER, M., HSU, C.C. and PEARLMAN, R. (1994) Feasibility study on spray-drying protein pharmaceuticals: Recombinant human growth hormone and tissue-type plasminogen activator. *Pharmaceut. Res.* **11**, 12–20.

NEWMAN, S.P. and CLARK, S.W. (1983) Therapeutic aerosols. I. Physical and practical considerations. *Thorax* **38**, 881–886.

OESWEIN, J.G. and PATTON, J. (1990) Aerosolization of proteins. In: *Proceedings of Respiratory Drug Delivery II, Keystone, CO.* March.

OTSUKA, M. and KANENIWA, N. (1990) Effect of grinding on the crystallinity and chemical stability in the solid state of cephalothin sodium. *Int. J. Pharm.* **62**, 65–73.

PHILLIPS, E.M. and STELLA, V.J. (1993) Rapid expansion from supercritical solutions: application to pharmaceutical processes. *Int. J. Pharm.* **94**, 1–10.

PODCZECK, F., NEWTON, J.A. and JAMES, M.B. (1996) The adhesion strength of particles of salmeterol base and salmeterol salts to various substrate materials. *J. Adhes. Sci. Technol.* **10**, 257–268.

RIETEMA, K. (1991) Theoretical derivatives of interparticulate forces. In: *The Dynamics of Fine Powders.* New York, Elsevier, 65–94.

SALEKI-GERHARDT, A., AHLNECK, C. and ZOGRAFI, G. (1994) Assessment of disorder in crystalline solids. *Int. J. Pharm.* **101**, 237–247.

SHEKUNOV, B.Y. and YORK, P. (2000) Crystallization processes in pharmaceutical technology and drug delivery design. *J. Cryst. Growth* **211**, 122–136.

STANIFORTH, J.N. (1982a) The effect of frictional charges on flow properties of direct compression tableting excipients. *Int. J. Pharm.* **11**, 109–117.

CHAPTER 5

STANIFORTH, J.N. (1982b) Relationship between vibration produced during powder handling and segregation of pharmaceutical powder mixes. *Int. J. Pharm.* **12**, 199–207.

STANIFORTH, J.N. (1985) Ordered mixing or spontaneous granulation? *Powder Technol.* **45**, 73–77.

STANIFORTH, J.N. (1994) The importance of electrostatic measurements in aerosol formulation and preformulation. In: BYRON, P.R., DALBY, R.N. and FARR, S.J. (eds), *Respiratory Drug Delivery IV, Program and Proceedings, Virginia, USA.* 303–311.

STANIFORTH, J.N. (1995) Performance-modifying influences in dry powder inhalation systems. *Aerosol Sci. Tech.* **22**, 346–353.

STANIFORTH, J.N. (1996) Improvement in dry powder inhaler performance: surface passivation effects. In: *Proceedings of Drug Delivery to the Lungs VII (London).* December, 86–89.

STANIFORTH, J.N. and REES, J.E. (1982) Electrostatic charge interactions in ordered powder mixes. *J. Pharm. Pharmacol.* **34**, 69–76.

STANIFORTH, J.N. and REES, J.E. (1983) Segregation of vibrated powder mixes containing different concentrations of fine potassium chloride and tablet excipients. *J. Pharm. Pharmacol.* **35**, 549–554.

SUTTON, H.M. (1976) Flow properties of powders and the role of surface character. In: PARFITT, G.D. and SING, K.S.W. (eds), *Characterization of Powder Surfaces with Special Reference to Pigments and Fillers.* London, Academic Press, 122.

TEE, S.K. (1996) An investigation on the effect of adherent fine particles in dry powder formulation for drug delivery to the lung. MSc thesis, University of London.

TIMBRELL, V. (1965) The inhalation of fibrous dusts. *Ann. NY Acad.* **132**, 255–273.

TIMSINA, M.P., MARTIN, G.P., MARRIOTT, C., GANDERTON, D. and YIANNESKIS, M. (1994) Drug delivery to the respiratory tract using dry powder inhalers. *Int. J. Pharm.* **101**, 1–13.

VENTHOYE, G., TAYLOR, K.M.G., NEWTON, J.M. and JOHNSON, M.C.R. (1995) The use of two inertial impaction methods to determine the effect of the size fraction of lactose on the fine particle dose delivered from mixes with salbutamol sulphate. In: *Drug Delivery to the Lungs VI, Proceedings, London.* 56–59.

VIDGREN, M.T., VIDGREN, P.A. and PARONEN, T.P. (1987) Comparison of physical and inhalation properties of spray-dried and mechanically micronised disodium cromoglycate. *Int. J. Pharm.* **35**, 139–144.

VIDGREN, M., VIDGREN, P., UOTILA, J. and PARONEN, P. (1988) *In vitro* inhalation of disodium cromoglycate powders using two dosage forms. *Acta Pharm. Fenn.* **97**, 187–195.

WARD, G.H. and SCHULTZ, R.K. (1995) Process-induced crystallinity changes of albuterol sulphate and its effect on powder physical stability. *Pharmaceut. Res.* **12**, 773–779.

ZENG, X.M. (1997) Influence of particle engineering on drug delivery by dry powder aerosols. PhD thesis, University of London.

ZENG, X.M., TEE, S.K., MARTIN, G.P. and MARRIOTT, C. (1996) Effects of mixing procedure and particle size distribution of carrier particles on the deposition of salbutamol sulphate from dry powder inhaler formulations. *Proceedings of Drug Delivery to the Lungs VII (London)*. December, 40–43.

ZENG, X.M., MARTIN, G.P., MARRIOTT, C. and PRITCHARD, J. (1997) Effect of surface smoothness of lactose on the delivery of salbutamol sulphate from dry powder inhalers. *Pharmaceut. Res.* **14**, S136.

ZENG, X.M., MARTIN, G.P., TEE, S. and MARRIOTT, C. (1998) The role of fine particle lactose on the dispersion and deaggregation of salbutamol sulphate in an air stream *in vitro*. *Int. J. Pharm.* **179**, 99–110.

ZENG, X.M., MARTIN, G.P., TEE, S., GHOUSH, A.A. and MARRIOTT, C. (1999) Effects of particle size and adding sequence of fine lactose on the deposition of salbutamol sulphate from a dry powder formulation. *Int. J. Pharm.* **182**, 133–144.

ZENG, X.M., MARTIN, G.P., MARRIOTT, C. and PRITCHARD, J. (2000a) The influence of crystallization conditions on the morphology of lactose intended for use as a carrier for dry powder aerosols. *J. Pharm. Pharmacol.* **52**, 633–643.

ZENG, X.M., MARTIN, G.P., MARRIOTT, C. and PRITCHARD, J. (2000b) The influence of the carrier morphology on the drug delivery by dry powder inhalers. *Int. J. Pharm.* **200**, 93–106.

ZENG, X.M., PANDHAL, K.H. and MARTIN, G.P. (2000c) The influence of fine lactose on the content homogeneity and dispersibility of beclomethasone diproprionate from dry powder aerosols. *Int. J. Pharm.* **197**, 41–52.

CHAPTER 5

Measurement of Cohesion, Adhesion and Dispersion of Powders

Contents

6.1 INTRODUCTION

Numerous attempts have been made to describe particulate interactions quantitatively, especially particle adhesion onto a surface (e.g. Corn, 1966; Krupp, 1967; Zimon, 1982). However, so far, practically no physical model that has been universally accepted to predict particulate interactions accurately, due to the complexity associated with these processes. It is not uncommon, for example, that particles from the same batch of powders may exhibit completely different adhesion characteristics to the same substrate surface with slight changes in ambient conditions, such as that produced by varying environmental relative humidity. It is rarely the case for all particles of the same test sample to show the same adhesion forces to a substrate surface due to differences in either the local surface texture or surface energy of the substrate. Therefore, adhesion forces between component particles in a powder are almost always described statistically by an adhesion force distribution, a typical example of which is shown in Figure 6.1 and it can be seen that particles exhibit a range of adhesion forces to a specific substrate surface. The median adhesion force, F_{50}, is often employed to quantify the overall adhesion. This value is by definition a single point and other parameters such as standard deviation, interquartile range, etc. have also been employed to describe more fully the distribution of forces.

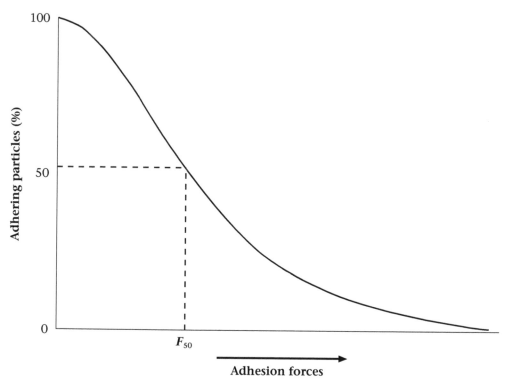

Figure 6.1: A typical adhesion force distribution of particles to a substrate surface.

All the mathematical equations generated to calculate particulate interactions are based upon certain assumptions (see chapter 1), e.g. a geometrically simple particle adhered to a perfectly smooth surface. It may be dangerous to extrapolate the theories to predict a real situation where morphologically complicated particles often interact with a substrate surface that has a range of surface asperities and surface free energies. Thus, although numerous such equations have been proposed particularly in colloid and surface science, experimental measurement is still essential in dealing with particulate interactions. Only results from the latter can provide an actual insight into the extent of particulate interactions.

6.2 COHESION OF POWDERS

There are several typical methods of measuring cohesive behaviour of ensembles of particles, most of which are qualitative rather than quantitative. These include measurement of:

- Tensile strength.
- Minimum diameter of an opening through which a particle ensemble is capable of passing without application of external forces, i.e. solely under its own weight.
- Dispersion of particle aggregates on impact after a fall from a given height.
- Dispersion of particles by an air jet.
- Shear strength of a mass of particles, in particular measurement of the internal friction by use of a rotating viscosimeter.
- Particle flow using an inclined plane whereby a horizontal plane covered with a particle layer is tilted until the layer starts to slide under gravity. The tilting angle is taken as a measure of the adhesion of the layer. This procedure may, in some cases, correspond to a cohesion measurement within the particle mass. If particles form a more or less tenaciously adhering layer on the substrate surface, particles in higher layers may slide upon tilting the substrate surface.

6.2.1 Angle of repose and the angle of internal friction

If a powder is poured (or cascaded) onto a flat surface, the powder will always arrange itself in a cone (Figure 6.2). When the gravitational force due to the powder mass is the only external force acting on the powder, the angle of the cone to the horizontal, the angle of repose, cannot exceed a certain value. If any particle temporarily lies outside the limiting angle, it will slide down under the force of gravity, until the free surface corresponding to the angle of repose is achieved, where the gravitational pull is balanced by the mutual friction and cohesive forces.

CHAPTER 6

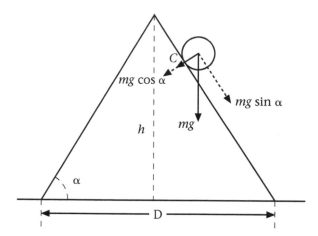

Figure 6.2: Schematic diagram of the effects of gravity and cohesive force on a particle lying on the slant of a heap.

The angle of repose can be calculated as:

$$\tan \alpha = \frac{2h}{D} \tag{6.1}$$

where h is the height and D is the diameter of the base of the powder cone.

As shown in Figure 6.2, the behaviour of a particle which is positioned on the slant of the cone is governed by two forces: (1) the gravitational force, which is vertical to the base of the cone, and (2) the cohesional forces, which are normal to the slant surface. The forces acting on the particle due to the gravitational force are composed of a force which is perpendicular ($mg \cos \alpha$) and one which is tangential ($mg \sin \alpha$) to the slant surface. Therefore, the normal force, τ, of the particle to the slant surface is the sum of the cohesional force and the part of gravitational force that is perpendicular to the slant surface:

$$\tau = mg \cos \alpha + C \tag{6.2}$$

where m is the particle mass, g is gravitational constant, α is the angle of repose and C is the cohesional force.

When these forces reach an equilibrium for a particle, then the following equation is obtained:

$$mg \sin \alpha = \mu(C + mg \cos \alpha) \tag{6.3}$$

where μ is the internal friction of the particles. Thus, the cohesional force can then be calculated as:

$$C = \frac{mg \sin \alpha}{\mu} - mg \cos \alpha \tag{6.4}$$

6.2.2 Shear measurement of powder

The cohesion and flow properties of powder packings may be defined more precisely as their shear strength or resistance to shearing under different normal loads, and this can be measured by different shear testers. The best known tester involves an apparatus based on a design by Jenike (1964). The Jenike shear-test cell consists of a cylindrical cell, split horizontally, to form a fixed base and a movable upper half (Figure 6.3).

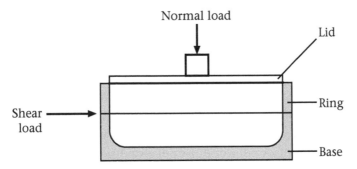

Figure 6.3: The Jenike shear-test cell.

The powder is packed carefully into the assembly and consolidated. A normal load is applied to the top of the powder specimen via the lid and the shear force is then applied horizontally to the movable half of the cell at a constant but low rate. Then, the shear stress needed to cause powder failure is measured. If the test is carried out for each of a series of normal loads on each prepared sample, then a plot of shear stress, τ, against the normal load, σ, which is called yield locus, can be obtained (Figure 6.4).

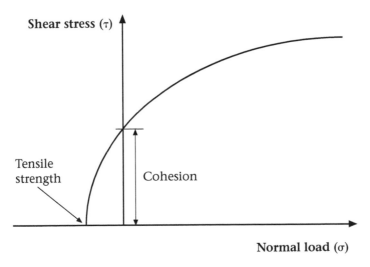

Figure 6.4: A typical yield locus for compact powders.

Certain terms are commonly used in powder rheology to refer to different points on the yield locus, such as cohesion and tensile strength. Cohesion is defined as the shear stress required for failure of a powder bed when zero load is applied normal to the shear plane (Figure 6.4) whilst tensile strength is the normal stress required for failure of the compact at zero shear stress.

6.2.3 Microbalance method

An apparatus based on a balance and spring may also be employed to measure the cohesive forces of bulk of powdered material. A schematic diagram of a typical such apparatus is shown in Figure 6.5 (Fukuoka *et al.*, 1983).

The apparatus illustrated in Figure 6.5 consists of a scale pan (about 50 mg in weight) hung from a thin spring connected to a strain gauge transducer with a nylon thread, a bridge circuit, an amplifier and a recorder. Before measurement, a thin layer of white petrolatum is spread on the bottom of the scale pan as an adhesive agent. The powder sample is placed on the jack, which is then elevated slowly until the powder surface comes into contact with the bottom of the scale pan. Different loads can be applied to the powder bed by lowering or raising the jack. After the application of the required

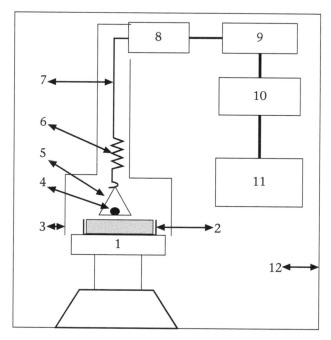

Figure 6.5: Apparatus used for the measurement of cohesiveness of powdered materials (adapted from Fukuoka *et al.*, 1983). (1) Jack; (2) powder sample; (3) windbreak; (4) weight; (5) scale pan; (6) thin spring; (7) nylon thread; (8) strain-gauge transducer; (9) bridge circuit; (10) amplifier; (11) recorder; (12) chamber of constant humidity and temperature.

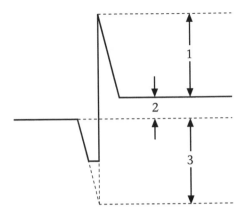

Figure 6.6: A typical curve showing the change in the readings of balance during a cycle of cohesiveness measurements. (1) Cohesive forces; (2) weight of the particle adhered to the scale pan; (3) load applied to the scale pan.

load for a period of approximately 1 s, the sample is lowered slowly until the adhered particles separate from the rest of the powder bed. The forces required, i.e. the sum of the tensile strength, C_0 and the weight of the particles adhering to the pan, can be obtained from a typical recorded signal as shown in Figure 6.6.

6.3 ADHESIVE FORCES

Adhesion forces can only be measured in a destructive manner and, therefore, an adhesion measurement does not represent an equilibrium situation so that a kinetic model of adhesion is required. Adhesion of a particle to a substrate surface involves the following essential steps. First, particle and substrate come into contact at one point by a contact area of atomic dimensions. Second, the long range attraction forces between the surface and particle bring the particle closer to the substrate such that the contact area between the two increases due to deformation of either the substrate or the particle. The contact area increases until the attractive forces and the forces resisting further deformation at the interface reach an equilibrium.

There are many methods that can be employed to measure the adhesion forces. These can be divided into two broad categories, indirect methods and direct methods. The adhesion forces of particles to a substrate can be obtained by calculation from indirect methods.

6.3.1 Tilted plane method

In the tilted plane method, a small amount of powder is placed on a flat surface such that a monolayer of powder is formed on the substrate surface, which is then tilted at an

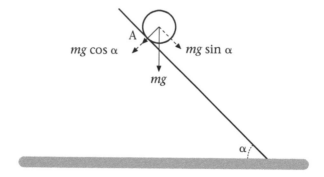

Figure 6.7: Various forces acting on a particle on a tilted plane.

ever-increasing angle until the powder slides off (Figure 6.7). This process can be described in a manner similar to that used in measuring the angle of repose.

When the tilting angle reaches a point such that a particle starts to slide off the plane, then the following relationship between the various forces acting on the particle is achieved:

$$mg \sin \alpha = \mu(mg \cos \alpha + A) \tag{6.5}$$

where m, g, α and μ are defined as before and A is the force of adhesion of the particle to the substrate plane.

The adhesion force of the particle to the plane surface can then be calculated by rearranging equation (6.5) to obtain:

$$A = \frac{mg \sin \alpha}{\mu} - mg \cos \alpha \tag{6.6}$$

Due to its simplicity, the tilted plane method can only be regarded as empirical and can only be expected to provide qualitative rather than quantitative information on the average adhesion forces of a powder to a plane surface. In order for the results obtained by the tilted plane method to represent the true adhesion situation, the powder must slide off the plane as a 'rigid solid'. However, the complicated nature of a sliding process makes the tilted plane method unsatisfactory for the measurement of the adhesion forces of particles to the plane surface. Sometimes the particles may adhere to the substrate so strongly that a shear plane develops in the powder bed, in which case the measured property is governed by the cohesion instead of the adhesion force. Further, particles may move individually. In either case, the results obtained by the tilted plane method cannot represent the actual adhesion of the powder to the plane surface. Moreover, in order for the adhesion to be representative of that of the bulk of the powder, the particle size, size distribution and particle shape at the powder–plane interface must be reproducible and representative of the properties of the bulk powder.

Thus, the tilted plane method often measures the combination of adhesion (particle–surface), cohesion (particle–particle) and frictional forces. It is therefore often difficult to interpret the results obtained in terms of a theoretical model of adhesion.

6.3.2   The gravity method

In the gravity method, the particle is placed on the substrate surface which is then moved continuously in the horizontal direction until it is separated from the substrate surface (Figure 6.8). The adhesion force of the particle to the surface is then equal to the gravity components that tend to separate the particle (Howe *et al.*, 1955):

$$A = F = mg \sin \varphi \tag{6.7}$$

where F is the separation force, m is the mass of the particle, g is the acceleration due to gravity and φ is the angle of inclination.

The gravity method can be employed to measure the adhesion force of comparatively large particles with diameters of the order of 1 mm.

■
CHAPTER 6
■

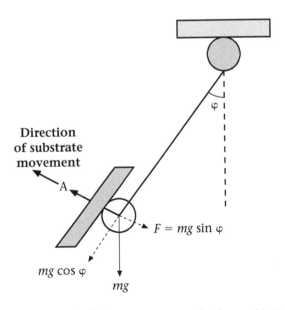

Figure 6.8: A schematic diagram of adhesion measurement by the gravity method.

6.3.3 Centrifuge method

The adhesive forces between particles and a substrate surface can be measured with satisfactory accuracy by centrifugation. This method is one of the most commonly employed techniques to measure adhesion forces of pharmaceutical powders (Lam and Newton, 1991; Podczeck *et al.*, 1995). In the centrifuge method, particles are applied to a substrate surface in such a manner that the centrifugal force on each particle acts either perpendicularly or parallel to the substrate. The surface covered with the particles is then subjected to centrifugation at an increasing angular velocity ω (Figure 6.9).

The force required to dislodge a spherical particle from a substrate by centrifugal action can be calculated from:

$$F = \frac{1}{6}\,\pi d_p^3 \rho_p \omega^2 l \qquad\qquad (6.8)$$

where F is the removal force; ω is the angular speed of rotation; d_p is the particle diameter; l is the distance from the centrifuge axis of rotation to the substrate; ρ_p is the particle density. The force of adhesion is equal in magnitude, but opposite in sign to the centrifugal force required to dislodge the particles from the substrate.

The substrate surface for adhesion measurements using a centrifugal technique usually comprised a flat disc made from metal, plastic or a compacted powder (tablet) to which individual particles are adhered. The discs have to be fixed at a defined position to the direction of the centrifugal force vector and at a defined distance between the surface and rotor centre. A special centrifuge tube was constructed by Booth and Newton (1987) to hold the test substrate (Figure 6.10). The substrate inside the centri-

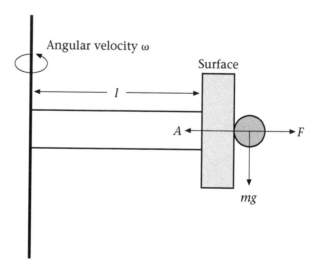

Figure 6.9: Adhesion measurement by centrifugation.

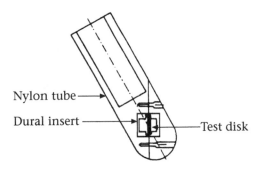

Figure 6.10: A diagram showing the centrifugal tube employed by Booth and Newton (1987) to measure adhesion forces.

fuge tube is orientated in such a way that the surface is parallel to the vertical axis of rotation of the centrifuge, and the acceleration forces act in a direction normal to the particle–substrate contact plane. Two different forces are applied to adhesion samples using the centrifuge technique: (1) a press-on force to enhance the contact between the particles and the surface prior to the determination of the number of particles adhered and (2) a systematically increased spin-off force to gradually detach the adhering particles. In this way, it is possible to obtain the adhesion force distribution of the particles to the surface (Podczeck *et al.*, 1994).

In reality, it is necessary to use a microscope to determine the number of particles of a given size adhering to the substrate before and after centrifugation at a given speed of rotation. The forces acting on particles of different sizes are then related to the efficiency of particle removal. The centrifugation technique enables investigators to make simultaneous measurements on a large number of particles.

After centrifuging at an initial small angular velocity ω, the centrifuge is stopped and by comparing the number of particles adhering to the substrate surface before and after the centrifugation, the number of particles which are dislodged from the surface can be calculated. This process is continued at increasing angular velocities until all particles are removed. Depending upon the particle size of the adherents and the specific measurement conditions, the adhesive forces measured by this method are approximately of the order of 10^{-6} to 10 N. However, the adhesion of very small particles can be evaluated by using an ultracentrifuge, where a force of up to $2 \times 10^5 g$ is exerted (Boehem *et al.*, 1962). With non-spherical particles, it is necessary to assume a particle shape factor and an associated characteristic diameter in order to calculate centrifugal force by means of equation (6.8).

During measurement, the powder material should be gently 'dusted' onto the substrate surface to ensure that the particles are separated from each other by an average distance larger than the particle diameter, in order to minimise the influence on adhesion due to interparticle interactions (Lam and Newton, 1991). The 'dusted' disc is then gently turned upside down to allow any loose particles to fall off. The prepared disc is subsequently mounted in the adhesion cell which is then fitted into a rotor boring. Initially, the disc is mounted with the 'dusted' surface facing towards the axis of rotation so that during centrifugation, the particles are forced onto the substrate surface by the centrifugal force. Then, the 'dusted' surface is placed facing away from the rotational axis such that centrifugation forces result in the detachment of particles from the substrate surface. The number of particles that remain after centrifugation is counted under an optical microscope and the adhesion profiles are obtained from the percentage of residual particles remaining on the disc after each centrifugation against the corresponding detaching forces. Each particular adhesive system can then be characterised by the geometric median adhesion force and the geometric standard deviation. The median adhesion force can be defined as the force of adhesion at which there is a 50% probability of the particles remaining adhered to the substrate after centrifugation, and this is used to represent the average adhesion force of a particle to the substrate. The geometric standard deviation is the ratio of the force required for 50% detachment of particles from the substrate to that required for 16% particle detachment and this is employed to characterise the distribution of the adhesion forces (Lam and Newton, 1991).

6.3.4 Microbalance methods

Microbalances, usually based on the use of quartz fibres which can take the form of a cantilever, helical spring or torsion balance can be employed in determining adhesion forces between particles. A microbalance technique requires that a single particle, preferably less than 50 μm in diameter, be attached to the balance spring or beam. Such an operation may be very difficult to control since apart from the particle and surface properties, particle adhesion is also determined by the microscopic contact between the particle and the substrate surface. The contact state between the particle and surface may vary during measurement and thus, a range of forces can be obtained for repeated contacts of the same particle on the same substrate (Corn, 1961). The measurement of the range of adhesive forces possible with particles of different sizes becomes a very time-consuming and tedious task.

6.4 AERODYNAMIC METHODS

Both the dispersibility in an air stream and the removal of adherents by an air stream can be employed to evaluate the cohesion and adhesion of particulate materials. The dispersion of particle aggregates by an air stream can be classified using the following techniques (Kousaka, 1991): (1) dispersion by acceleration flow or shear flow; (2) dispersion by impaction of aggregates onto targets and (3) dispersion by mechanical forces such as fluidisation and vibration. The most common method of dispersion is to feed the powder into a high-velocity air stream with the dispersed particles being collected onto a surface, e.g. a glass slide. The particle size distribution of the collected particles is then examined using an optical or electron microscope. By comparing the particle size distribution of the sample before and after dispersion, the dispersibility of the powder by the air stream can be obtained. Alternatively, the particle size distribution of dispersed particles can be analysed *in situ* by techniques such as laser light diffraction. The drag forces of the air stream can then be calculated under defined conditions. If the particle size and the state of aggregation (i.e. the number of primary particles contained in each aggregate) are known, then the cohesion forces between these particles can be calculated (Endo *et al.*, 1997).

Similarly, particle adhesion to a surface can also be measured by means of an air stream. When an adherent particle is exposed to a compressed air stream, then it will be dislodged from the adhering surface if the drag forces of the stream overcome the adhesional forces of the particle.

The limitation of this method resides in the difficulty of defining the aerodynamic conditions in the vicinity of the dispersed particles. Due to the lack of a well-defined velocity profile of the dispersing air stream, the results obtained with the method must be expressed as efficiency of particle removal or deaggregation versus the air velocity characteristics such as the linear velocity, acceleration and duration of the air flow treatment. However, it is possible to estimate the adhesive and/or cohesive forces of particulate materials in a well-defined velocity field.

6.4.1   Measurement of powder dispersion

Due to the small size of the powder particles in an aerosol, when such particles collide with one another, they are likely to cohere to form larger agglomerates with more rapid settling tendencies. Also the collision of particles with the walls of a dispersing container may result in adhesion to these surfaces, leading to their slow but sure extraction from the system.

Dispersion is often carried out by passing a metered quantity of the particulate material with a gas stream through a nozzle. Often this is nothing more than a round tube where high velocities, hence considerable shear and attrition, are developed to tear the particles one from another. Figure 6.11 shows a schematic diagram of an experimental apparatus employed for the measurement of particle adhesion on a surface using this technique (Gotoh *et al.*, 1994). During operation, compressed air is generated from the compressor and is then adjusted to an appropriate pressure before it reaches the jet nozzle, which is a rectangle with dimensions of $0.25 \times 10 \, mm^2$. The tip of the nozzle is placed 10 mm above the test sample, which has been previously sprayed onto a

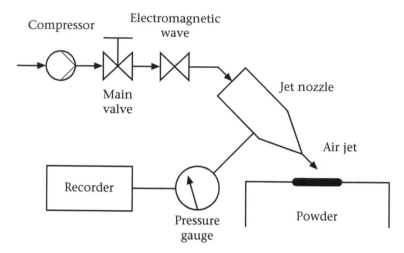

Figure 6.11: A schematic diagram showing the jet nozzle method employed in the measurement of particle adhesion to a surface (Gotoh *et al.*, 1994).

substrate such as glass. After having been subjected to the compressed air for a predetermined period of time from a specific jet-impinging angle, the particles remaining on the substrate surface are counted using a light microscope. Comparison of the number of particles before and after treatment, enables the particle removal efficiency to be quantified, and this in turn can be employed to evaluate the adhesion forces of the powder to the substrate surface.

Obviously, the air velocity, turbulence and particle interaction within the powders and the duration of dispersion are among the most important factors that determine the extent of powder removal from the adhering surface. For example, the longer the tube and the smaller its diameter the more complete the dispersion at a given gas velocity. However, for a specific system, if the instrumental factors and the properties of the dispersing air stream are kept constant (or are known), then powder dispersion is mainly dependent upon particulate interactions, i.e. higher particulate interaction produces lower dispersion. While attempts have been made to describe dispersion phenomena mathematically, the wide variation in the adhesiveness of powders has not yet allowed a general interpretation. But with the measurement of dispersion of a powder under controlled conditions, it is possible to estimate such forces in a powder bed.

6.4.2 Sampling of airborne particles

There are two basic approaches to the measurement of airborne particles. The traditional approach is to sample the particles on a collection surface followed by analysis of those collected. The other approach is to sample the particles directly into a real-time, dynamic measuring instrument. In the first method, the particles are collected for measurement at a later time, and a range of analytical techniques may be used. One of the drawbacks of this method is the long time delay between sampling and the subsequent analysis. In contrast, real-time measurement methods provide rapid data since many direct reading instruments yield instantaneous, or virtually instantaneous information, on particle concentration and size distribution.

Filtration is the most widely used method for removing particles from an air stream for subsequent analysis (Lippmann, 1989). Particle collection in a filter involves many mechanisms, such as inertial impaction, interception, diffusion, gravitational settling and electrostatic attraction (Hinds, 1982). The relative importance of these different mechanisms depends on the size, density, shape, and electrical charge of the collected particles, the filtration flow velocity and the mechanical and electrical properties of the filter.

Inertia is an important particle property that is utilised to remove particles from an air stream. If the flow of air changes direction, particle inertia will make the particle trajectories deviate from the flow streamlines. In an impactor, the particle-laden air flow is pulled or pushed through a nozzle, which accelerates the air and the particles to high speed. The air jet emerging from the nozzle is deflected by the impaction surface and this induces the air to make a sharp turn. Particles with sufficient inertia continue their straightforward movement towards the collection surface and impact onto the collection surface (Marple and Willeke, 1979). An impaction stage may be preceded by a single nozzle or many nozzles next to each other. By combining several single-stage impactors into a multistage or cascade impactor, the particles can be collected in a size-selective manner and each stage represents a specific particle size range. This is achieved by decreasing the nozzle diameter, the number of nozzles, or both, in each successive stage. By so doing, the linear air velocity will increase in order to keep the volumetric flow constant and thus, particle inertia is increased progressively. If a small particle fails to have sufficient inertia to deposit at a particular stage, it may deposit at a later stage due to increased particle inertia at that stage. In operation, each stage is assumed to collect all particles of a diameter larger than its cut-off size. As a consequence of particle collection on the previous stage, the upper limit of particle size captured at each given stage takes the value of the cut-off size of the previous stage. Thus, each stage collects particles with diameters in a limited size range defined by successive cut-off particle sizes.

In the inertial impaction technique, particles may be impacted onto a stagnant volume of air instead of a solid surface. Particles of high inertia move across the virtual surface between the air flow and the stagnant air volume and are collected in the latter, whilst the small particles are deflected at the virtual interface in a manner similar to solid–surface impaction. The airflow itself and the impaction of large particles will cause some turbulence in the otherwise stagnant air volume, which may return some of the large particles back into the air stream. Hence, the air beyond the virtual interface is withdrawn at a slow rate to collect particles, making this type of impactor a two-flow or 'dichotomous' impactor. In both types of flow, separated particles can be collected subsequently on a filter or analysed in a real-time instrument.

Another mechanism of particle collection is the use of a cyclone in which particles are moved by centrifugal forces. In a cyclone collector, the particle-laden air stream moves in a helical path. Particles with high inertia are separated by the centrifugal force whilst smaller particles continue with the air flow. In general, a cyclone does not yield as sharp a cut-off in particle size as an impactor. However, a cyclone separator has a high loading capacity when compared to that of an impaction stage. Thus, it is a good choice when large numbers of particles are to be sampled.

In the case of inertial separation, the particle density, size and velocity are principal parameters contributing to the driving force for particle separation from the air flow. In fact, the gravitational force acting on the particles can be employed to separate them into different size ranges. An instrument that employs gravitational forces to collect particles in a size-selective manner is called an elutriator. The particles are either removed or elutriated by the relative velocity between the air flow and the gravitational settling of the particles (Hinds, 1982). A vertical elutriator consists of a vertical tube or channel in which the air stream flows upwards at low velocity. Only particles with a gravitational-settling velocity that is less than the upward air velocity remain airborne whilst particles that have a gravitational-settling velocity higher than the air velocity will deposit in the device. A horizontal elutriator consists of a horizontal duct or a set of parallel ducts of rectangular cross-section. Separation of particles with different sizes is due to size-dependent settling in the gravitational field. The major advantages of the horizontal elutriator are its simplicity of construction, ease of modelling the size separation mechanism and high loading capacity. In general, the horizontal elutriator is used to collect a sample of large particles or to separate large particles from smaller ones.

Gravitational forces may not be sufficiently efficient to cause the settlement of small particles. If this is the case, centrifugal force can be applied to separate particles of different sizes (Tillery, 1979). Obviously, when centrifugal force is much higher than gravitational force, it is possible to extend the measurements to lower particle size ranges.

6.4.3   Sample analysis

Once a sample of dispersed particles has been collected, the sample may be analysed as a whole or as individual particles. The method of analysis can be varied according to the nature of the particle.

6.4.3.1   Physical analysis

Gravimetry is one of the most widely used methods for the analysis of samples of aerosol clouds that have been collected. Conventionally, a known volume of dispersing air is sampled on a filter or alternative collection surface, which has been previously weighed. The mass increase of the collection surface is then divided by the air volume and the ratio is the concentration of the dispersed powder in the air stream. Obviously, careful control of the weighing process is crucial in maintaining the accuracy of the method and also, the effects of water vapour uptake by the sample and electrostatic charge of the sample particles on the balance reading, have to be minimised. In order to avoid the effect of static electricity, the electric charge on the collection surface and the powdered material is usually reduced by a charge neutraliser. The relative humidity of the environment is also an important factor to be considered. The major drawback associated with the use of gravimetry is its inability to reveal the particle size of the dispersed materials and hence, it is often unable to provide sound data on the polydispersibility of a powder.

Optical microscopy is one of the most important methods of characterising particle size, size distribution and morphology. The dispersed particles can be collected on a glass slide before undergoing microscopic analysis. Such a method is able to provide information on the size and size distribution of the dispersed particles. A conventional bright-field microscope can be used to count particles down to approximately 0.3–0.5 μm. A number of modifications can be made to the optical microscope to improve visualisation of the particles and these include the use of dark-field, polarising, phase-contrast, interference, fluorescence or ultraviolet illumination (Chung, 1981).

An electron microscope can measure ultrafine particles due to the much shorter wavelength of the radiation employed in this method. A scanning electron microscope can measure particles down to approximately 0.01 μm whilst with a transmission electron microscope, resolution can be as low as <0.001 μm. Since electron microscopic analysis is carried out in vacuum, this method is not suitable for materials that evaporate or degrade when exposed to a vacuum and/or heating by the electron beam.

6.4.3.2 Chemical analysis

Particles on collection plates may also be estimated by chemical analysis. Pharmaceutical powders may be composed of materials of varying chemical composition and, consequently, many analytical methods may be employed. The methods employed for chemical analysis are gas and liquid chromatography. However, powder beds can also be characterised directly by means of elemental analysis using atomic absorption spectroscopy, X-ray fluorescence analysis, proton induced X-ray analysis and neutron activation (Chung, 1981).

Atomic absorption spectroscopy is one of the most common methods for elemental analysis of dispersed powders. A typical procedure involves the preparation of a solution, which is then nebulised and vaporised by heating in the airborne state. The

CHAPTER 6

amount of each element is determined by measuring the light absorbance in the vapour at various wavelengths specific to each element. One of the major advantages of this method is its high sensitivity.

X-ray fluorescence spectrometry can be used to analyse elements of atomic number 11 (Na) and higher. In this method, the sample is bombarded with X-rays and the composite elements are excited emitting X-rays that are specific to each. The intensities of the element-specific X-rays are used to quantify the concentration in the powder.

Proton-induced X-ray emission analysis is based on a similar principle to X-ray fluorescence. However, the former utilises high-energy protons for excitation of the sample element and is more sensitive than conventional X-ray fluorescence. The sample elements may also be activated by neutrons, hence an alternative method termed neutron activation analysis can also be employed. Neutron bombardment creates radioactive isotopes which emit characteristic gamma spectra after activation which can subsequently be used to identify and quantify the elements in the sample.

6.5 CHARACTERISATION OF AERODYNAMIC PROPERTIES OF MEDICINAL AEROSOLS

The major purpose of formulating a drug as a dry powder for inhalation is to deliver the medicament exclusively to the site of action so as to maximise its therapeutic effect and to reduce any possible side-effects. Thus, the amount or percentage of the drug reaching these sites, which may be located in the peripheral airways, is one of the major factors to be considered in the design of a successful dry powder formulation. As discussed previously, the most important physico-chemical parameter influencing the deposition of aerosols in the lung is the particle or droplet size. However, it has to be noted that the aerodynamic diameter rather than the geometric diameter is employed to characterise the size of drug particles for inhalation. The former is defined as the diameter of a unit density sphere having the same settling velocity, generally in air, as the particle. It is an indicator of the aerodynamic behaviour of the particle and is dictated by particle shape, density, and geometric size, all of which can affect the deposition of the drug in the lung.

6.5.1 Inertial impaction

Inertial impaction is the most commonly used methodology to determine the aerodynamic particle size of aerosols. The devices employing inertial impaction to sample and determine the size fraction of airborne particles are thus called inertial impactors. The principle on which inertial impactors operate is shown in Figure 6.12. When the air flow changes direction, for example to go around a bend as shown in Figure 6.12, the airborne particles will be subjected to two forces. The first one is the momentum built up

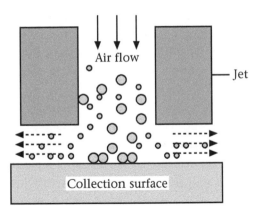

Figure 6.12: Schematic diagram of the inertial impaction of airborne particles.

whilst travelling along the streamline; the second one is the hydrodynamic force (friction) with the surrounding air stream as the latter changes direction. The momentum of the particles keeps them travelling in their established trajectory whilst the friction created between the particles and air causes the particles to accelerate in the new direction of the gas flow. Therefore, an airborne particle will continue to move in the original direction of flow until it loses inertia, it will then 'relax' into the new direction of flow with the time taken for this to occur known as the 'relaxation time'. Whether or not a particle will deposit on a collection surface placed in the path of the original direction of the air stream is determined by this relaxation time, which is in turn dependent upon particle momentum and the hydrodynamic force of the flow acting on the particle.

The possibility of a particle impacting on a surface is a function of a parameter known as the Stokes number (Hinds, 1982):

$$\text{St}k = \frac{\rho C V D_\text{p}^2}{9\mu D_\text{j}} \tag{6.9}$$

where Stk is the Stokes number; ρ is the particle density in g cm^{-3}; C is the Cunningham slip correction factor; V is the velocity in cm s^{-1}; D_p is the particle diameter in cm; μ is the fluid viscosity in poise; D_j is the jet diameter in cm.

The Cunningham slip factor is given by:

$$C = 1 + \frac{1.6 \times 10^{-5}}{D_\text{p}} \tag{6.10}$$

The slip factor represents the ability of the particles to slip between the gas molecules as the particle size is reduced below the mean free path in the gas.

The value of Stokes number varies within the range 0 to 1. The higher the value of Stokes number, the more likely a particle is to deposit on a plane surface. A value for Stokes number over 0.2 is normally thought to bring about inertial impaction of the

CHAPTER 6

particles although there is no clear cut-off between particles which are deposited and those which are not. It is a common practice to quote the Stokes number that leads to a 50% probability of deposition, Stk_{50}. If the value of Stk_{50} is known, then the value of D_{50}, corresponding to the cut-off size of the impactor can be calculated from:

$$D_{50} = \sqrt{\frac{9\mu D_j}{\rho C V}} \sqrt{Stk_{50}} \qquad\qquad (6.11)$$

where all the variables are defined as above.

According to equation (6.11), the inertial impaction of a particle is a function of particle diameter and density; as well as the velocity and viscosity of the air stream and the diameter of the jet in which the stream originally flows before changing its direction.

Larger particle diameters and/or higher flow rates produce higher values of Stokes number and, thus, increase the possibility of particle deposition. This can be easily understood since particle size, density and velocity are the major components of momentum. So increasing any of the factors will increase particle momentum. Particles with higher momentum are less likely to relax into the new direction of the flow and hence are more likely to impact on the collection surface. On the other hand, smaller particles travelling at a lower flow rate have lower momentum. These particles will relax more quickly into the new flow direction than the larger ones and, therefore, will not encounter the collection surface. The likelihood of particle deposition is also decreased by increasing the viscosity of the air stream. This is because fluid viscosity is one of the main factors determining the drag forces of a stream acting on a particle, higher viscosity producing stronger drag forces on a particle (see chapter 4). Therefore, a fluid with higher viscosity will make it more likely for the suspended particles to flow along the fluid direction and, therefore, decrease the possibility of particle deposition.

A number of devices used to measure the aerodynamic diameter of airborne particles have been described in the literature, mostly concerned with the measurement of particulate pollution. From the pharmaceutical viewpoint the most relevant aerodynamic particle sizers are the liquid impingers (e.g. twin impingers, multistage liquid impingers) and cascade impactors (e.g. the Andersen sampler), which are widely used for the study of inhalation aerosols.

6.5.2. Two-stage liquid impinger

The two-stage liquid impinger, also called the twin impinger, is a two-stage cascade impinger which was developed specifically to assess the delivery of drugs from MDIs (Hallworth and Westmoreland, 1987). It was the first device operating on inertial impaction to be adopted by the British Pharmacopoeia (BP) (British Pharmacopoeia Commission, 1993). As depicted in Figure 6.13, the twin impinger contains two impaction stages, the upper one (the round bottom flask) operates at low velocity and

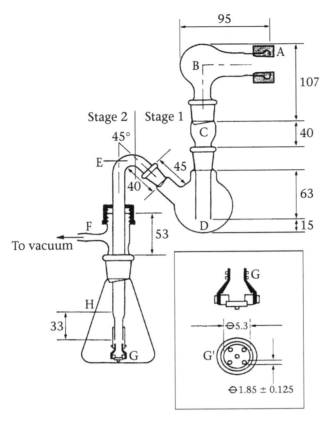

Figure 6.13: A schematic diagram of a twin-stage liquid impinger (reproduced from the British Pharmacopoeia, with permission).

traps the large particles whilst the lower stage (the conical flask) operates at a much higher velocity. The BP monograph provides a rigid dimensional specification for each component of the device. The twin impinger is designed to be a simple model of the respiratory tract, with the upper chamber representing the upper airways and the lower chamber representing the lower airways.

To operate the device, air is drawn through the instrument at a flow rate of $60 \, l \, min^{-1}$ through a pump connected to the outlet of the lower chamber. The aerosol cloud is introduced at the inlet and passes through a glass tube containing a 90° turn which is designed to simulate the oropharynx. It then passes into the upper impinger stage, which contains 7 ml of liquid (methanol : water mixture 50 : 50). Particles larger than the cut-off point of this first stage ($>6.4 \, \mu m$) should ideally be collected in the liquid. Smaller particles ($<6.4 \, \mu m$) that are not collected will proceed to the lower stage, which contains 30 ml of the same liquid. Most of these particles will be collected in this stage due to the excess of liquid and the much higher linear velocity of the air stream than in the upper stage. However, particles that are too small to be collected in the lower stage will be emitted at the exit. The amount of drug collected in the upper and lower stages can then be determined and particle size can be divided into two size ranges, i.e. particles larger than the cut-off diameter of stage 1 and particles smaller than this cut-off diameter.

The cut-off diameter of the upper stage at $60 \, l \, min^{-1}$ was determined to be 6.4 μm (Hallworth *et al.*, 1978; Miller *et al.*, 1992). The particle fraction that can reach the lower stage of a twin impinger has been defined as the 'respirable fraction', i.e. the fraction that is likely to penetrate deep into the lung and it has been reported to correlate with the clinical performance of a bronchodilator aerosol (Padfield *et al.*, 1983; Meakin and Stroud, 1983). However, the respirable fraction measured *in vitro* invariably overestimates the percentage of the emitted drug dose which is actually deposited in the lungs *in vivo* (Vidgren *et al.*, 1991). This discrepancy may be partly due to the complicated mechanisms of particle deposition in the lung and partly due to the oversimplification of the design of the impinger. Therefore, the particle fraction collected in the lower stage of a twin impinger is now referred to as the 'fine particle fraction' which reflects more accurately the physical means of particle capture.

A typical curve of the collection efficiency of particles in the lower stage of a twin impinger shows an 'S' shape (Figure 6.14). An ideal situation should involve a sharp division of particles into different size ranges, i.e. all particles that are larger than the cut-off point should be retained by the upper stage whilst those smaller than the cut-off point should reach the lower stage of the impinger. Departure from the ideal often results from the particle flow behaviour in areas near the impinger walls. The cut-off point is defined as the point of 50% collection efficiency and this is termed the effective cut-off diameter (ECD) (Hallworth and Westmoreland, 1987). Therefore, the division between the two stages of the impinger is not sharp, some of the data corresponding to particles collected in the lower stage will include a proportion larger than the nominal cut-off size (6.4 μm) whilst particles smaller than the cut-off value may not reach the lower stage. This broad division of particle size is one of the major problems associated with the use of the twin impinger in the characterisation of aerosol particle size.

Another limitation of the device arises from the fact that the twin impinger is merely a dichotomous sampler, i.e. the total sample is divided into only two size categories. It cannot reveal the particle size distribution of the aerosol cloud which is critical in predicting *in vivo* deposition. Particles having different sizes and/or size distributions may produce the same fine particle fractions when measured by a twin impinger. In theory, the number of combinations of powders with a particle size distribution that can give a specific fine particle fraction is infinite. In other words, all clouds composed of the same percentage of particles <6.4 μm will exhibit the same fine particle fraction. Examples of powders with different mass median aerodynamic diameter (MMAD) that can produce a specific fine particle fraction are listed in Table 6.1.

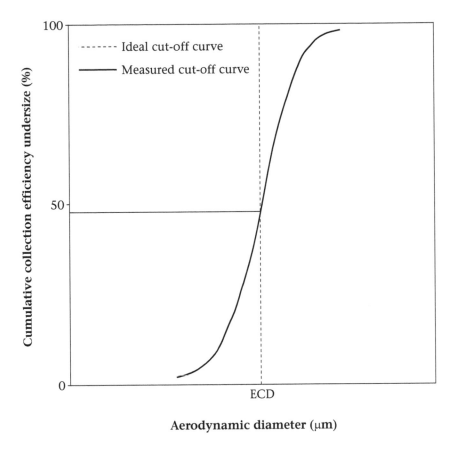

Figure 6.14: A typical collection curve of an impinger.

It can be seen that the same FPF may be obtained for particles of different size and size distribution. For example, particles with MMAD 7.0 and GSD 1.2 μm would give the same FPF of 30% as particles having MMAD 14.1 μm and GSD 4.5 μm. Any particles that have an MMAD of 6.4 μm would produce the same FPF value of 50% regardless of their size distribution. A similar calculation has been carried out by other workers (Miller *et al.*, 1992). Thus, the device is not able to discriminate among distributions with specific combinations of varying size and width. Similar FPF values obtained for different formulations may not indicate the equality of *in vivo* deposition after inhalation of these formulations since the lung deposition is a factor of both particle size and size distribution (Martonen and Katz, 1993). For example, a study using the twin impinger failed to show any differences between 12 different branded or generic salbutamol MDIs, although statistically significant differences between these MDIs were observed both in terms of the delivered dose and *in vivo* deposition of the drug (Lee *et al.*, 1993).

■ CHAPTER 6 ■

TABLE 6.1

Powders with different mass median aerodynamic diameter (MMAD) and geometric standard deviation (GSD) that produce the same fine particle fraction (FPF, <6.4 μm)

Aerosol	FPF = 30%		FPF = 40%		FPF = 50%	
	MMAD	GSD	MMAD	GSD	MMAD	GSD
1	14.1	4.5	8.8	3.6	6.4	3.1
2	12.1	3.4	8.3	2.8	6.4	2.5
3	10.1	2.4	7.7	2.1	6.4	1.9
4	8.7	1.8	7.3	1.6	6.4	1.6
5	7.7	1.4	6.9	1.4	6.4	1.3
6	7.0	1.2	6.6	1.2	6.4	1.1

(Data from Martonen et al., 1992.)

Although the twin impinger has the reputation of being a robust instrument, with small changes in operating parameters having little effect on the results, reports have shown that environmental conditions such as relative humidity and temperature can affect the calculation of both the emitted dose and FPF of dry powders (Jashnani et al., 1995). Increasing the relative humidity of the ambient atmosphere was shown to decrease the value obtained for the emitted dose and FPF of both salbutamol base and sulphate at any given temperature with differences being more marked at higher temperatures. These findings, if confirmed, could have great importance for the characterisation of aerosol dry powders since the BP test makes no mention of temperature and relative humidity of the ambient atmosphere. Therefore, the test conditions should be validated so as to obtain more reliable and reproducible results of aerosol deposition.

6.5.3 Multistage liquid impinger

Since the twin impinger is unable to measure the distribution of particle size of an aerosol, the multistage liquid impinger (MLI), which is based on a similar principle, viz the collection of airborne particles, was developed to cope with this problem. One of the earliest MLIs was a non-commercial four-stage glass liquid impinger (Figure 6.15) (Hallworth, 1977). The impaction surfaces consist of sintered-glass discs immersed in liquid, backed up by a liquid swirl impinger and a filter to trap very fine particles. Twenty millilitres of liquid is introduced into each stage such that the entire surface of all the glass plates are wetted through capillary action. The cut-off diameters for stages 1, 2, 3 and 4 are 10.5, 5.5, 3.3 and 0.8 μm respectively at a inhalation flow rate of 60 l min^{-1} (Jaegfeldt et al., 1987). Therefore, particles collected in stages 2–4 can be considered to be the respirable fraction.

Although the glass four-stage liquid impinger is no longer in production, MLIs of four

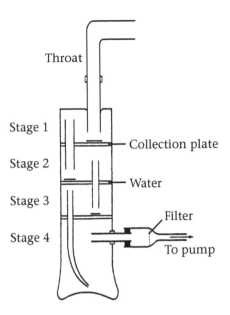

Figure 6.15: A schematic diagram of a glass four-stage liquid impinger (after Hallworth, 1977).

and five stages made from glass and metal have been commercialised and are widely employed in the characterisation of particle size of aerosols including dry powders. Both pieces of apparatus consist of a glass inlet throat (A), three impaction stages (for the four-stage impinger, Figure 6.16) or four impaction stages (for the five-stage impinger) and an integral filter stage. Each stage consists of a single jet and a sintered-glass plate. To operate this device, a suitable filter should be placed in the filter stage and 20 ml of a suitable solvent is introduced to each of the impaction stages. Powder deposition is tested at $60\,l\,min^{-1}$ for 5 s. Then, each stage is washed thoroughly with a suitable solvent and the amount of drug deposited is determined.

The cut-off diameters at $60\,l\,min^{-1}$ for stages 1, 2, 3 and 4 are approximately 12, 6.8, 3.1 and 1.7 μm, respectively and stage 5 is designed to capture the particles that have a diameter less than 1.7 μm (Hugosson *et al.*, 1993). The respirable dose is the total mass of the drug found in the relevant impactor stages (stages 3, 4 and 5). By constructing a curve of cumulative percentage undersize versus particle size, it is possible to estimate the mass median aerodynamic diameter and size distribution of the aerosol clouds.

It is apparent that the MLI has advantages over the TSI in that the former device can measure the particle size distribution, which is critical in respect of the deposition of particles *in vivo*. Further, MLIs have been calibrated to measure the particle size of an aerosol at flow rates ranging from 30 to $100\,l\,min^{-1}$ (Table 6.2). The effective cut-off diameter at 50% efficiency is inversely proportional to the square root of the flow rate

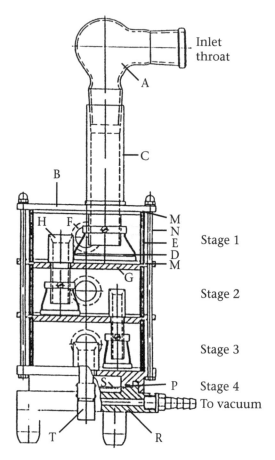

Figure 6.16: A schematic diagram of the commercial four-stage liquid impinger (reproduced from the British Pharmacopoeia with permission).

(Hugosson *et al.*, 1993). Thus, MLIs may be used to test aerosols at any flow rate within this range will undoubtedly expand the applicability of these devices since most other inertial impactors only operate at a single flow rate. It has long been observed that both inhaler devices and patients will generate different inhalation flow rates. The ability of MLIs to characterise aerosols at different flow rates will make it possible to simulate accurately the real situation of particle deaggregation on inhalation. Furthermore, MLIs are robust and reproducible and experiments carried out using such devices are not so time-consuming, as those which employ the cascade impactor (see section 6.5.4). There is no loss of drug particles on the wall of the impinger, which is commonly encountered with cascade impactors and thus MLIs may have distinctive advantages over other impactor devices.

TABLE 6.2
Nominal cut-off diameters of each stage of the MLI at flow rates of 30–100 l min^{-1}

Stage number	60 l min^{-1}	Other flow rates Q (l min^{-1})
Stage 1	13	13 $\times (Q/60)^{-1/2}$
Stage 2	6.8	6.8 $\times (Q/60)^{-1/2}$
Stage 3	3.1	3.1 $\times (Q/60)^{-1/2}$
Stage 4	1.7	1.7 $\times (Q/60)^{-1/2}$

6.5.4   The Andersen cascade impactor

The Andersen cascade impactor (ACI) is widely used in pharmaceutical aerosol characterisation. It was specifically designed to cover the size range of importance in pulmonary deposition, i.e. from 0.1 to 10 μm, which is well documented in terms of the deposition profiles in the lung. A schematic diagram of an Andersen cascade impactor is shown in Figure 6.17. As can be seen, this instrument consists of a stack of plates, akin to a stack of sieves. Each plate carries an array of finely drilled holes (400 per stage). The holes in the top plate are the largest, with diameters of 1.5 mm, and successive plates bear holes of decreasing diameter, the lowest being 0.25 mm. The final plate is unperforated and collects the finest particles. The diameter of the holes are chosen so that the air flow velocity increases down the stack, with the upper stages having the lowest linear velocity, and successive stages having an increasing velocity. As a particle passes down the impactor, its momentum increases, due to an increase in the linear velocity of the air stream, until it deposits on an impaction plate. In each stage, below the perforated plates are circular impaction plates on which the particles that have passed through the holes are deposited. The base of the impactor is connected to a vacuum pump, which draws air through the impactor. A glass throat piece is usually connected to the top of the impactor.

To operate the impactor, 10 ml of a suitable solvent (usually methanol–water mixture 50 : 50) is introduced to the pre-separator stage. After the device is found to be air tight and the air flow rate has been set at the desired level (28.3 l min^{-1} for the Andersen Mark II cascade impactor), the pump is switched on. The aerosol is then sprayed into the throat piece and after the pump has been allowed to run for 7 s, the impactor is dismantled and each stage is washed with a suitable solvent for further analysis.

The Andersen cascade impactor fractionates particles in the size range from 0.3 to 10 μm aerodynamic diameter, with nominal cut-off diameters of 9.0, 5.8, 4.7, 3.3, 2.1, 1.1, 0.7, 0.4 and 0.2 μm at a flow rate of 28.3 l min^{-1}. However, this impactor has also been calibrated at other flow rates. The cut-off diameter for each stage was shown to follow Stokes law, i.e. the cut-off diameter for a particular stage is inversely proportional to the square root of the air flow. Thus, the effective cut-off diameters of an ACI at a flow rate of 60 l min^{-1} are 6.18, 4.00, 3.20, 1.40, 0.76, 0.45 and 0.30 μm for stages 0, 1,

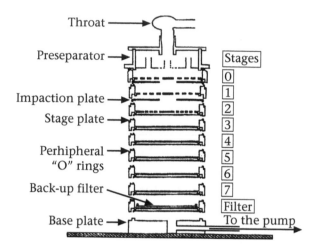

Figure 6.17: A schematic diagram of the Andersen cascade impactor.

2, 3, 4, 5, 6 and 7, respectively. This impactor is superior to both the twin impinger and multistage liquid impinger in that the ACI can produce a more detailed particle size distribution of the aerosolised drug. This is the prime advantage of the ACI, since, as has been stated, particle size distribution of the aerosolised drug particles is crucial in determining its deposition in the respiratory tract. However, such impactors have limitations and the three major areas of concern involve the problems of particle bounce from the collection surface, overloading of collected particle deposits on the impaction plates, and interstage losses.

Of these limitations, particle bounce has received the most attention, since particles bouncing from an impaction plate may be collected on any of the subsequent stages that have a smaller cut-off diameter, resulting in a biased measurement. The most widely used approach to solving a particle bounce problem is to coat the impaction plate with an appropriate material to convert it into a sticky surface. The stability, chemical composition and purity are all factors to be considered in selecting an appropriate substance. Numerous types of greases and oils, including petroleum jelly, Apiezon® greases, and silicone oils have been used to coat the impaction plates. If the impaction plate just has a sticky surface, once a monolayer of collected particles has formed on this surface, additional incoming particles will impact on the particles that have already been collected and the phenomenon of particle bounce is again possible. Therefore, it is preferable to use a sticky substance that will wick up through the particle deposit by capillary action and continually provide incoming particles with a sticky surface. Silicone oil works well in this respect.

A further error is associated with overloading if too much powder is deposited on any one of the plates. If the plate is examined under the microscope, the particles can be seen as small spots on the plate, beneath the corresponding jet. If the piles become too large, they will alter the impaction characteristics of the surface, and the particles may then be swept away by the air flow, leading to the results being biased towards the smaller size fractions. Such a phenomenon is called re-entrainment and can cause a significant measurement error. Overloading is an easier problem to solve since this can be simply tackled by reducing the volume of the aerosol cloud to be measured. The upper limit for the amount of a material that can be collected on each plate of the Andersen cascade impactor is 10 mg, which is usually considerably higher than the dose of drug required to be delivered *in vivo*.

The final area of concern is the collection of particles on surfaces other than the impaction plate, the so-called interstage loss. These particles are deposited on the interior surfaces of the impactor and are a function of particle size. In the upper stages, where the particles are relatively large, particles can be lost by impaction or turbulent deposition. As the particles travel down to the lower stages, the large particles will be removed and size of the airborne particles decreases successively. The interstage loss will also decrease until the particle size diminishes to such an extent ($<0.1 \, \mu m$) that diffusional deposition becomes significantly important. When the particles exceed $10 \, \mu m$ in aerodynamic diameter, interstage losses may increase above 20% of the total mass collected by the impactor (Mitchell *et al.*, 1988). Nevertheless, the effective cut-off diameters and the shapes of collection efficiency curves are not altered greatly. Furthermore, temperature and humidity will certainly have an effect on interstage losses in different stages of the impactor. One difficulty in estimating the interstage losses is that they vary with the nature of the particles. For example, if the particles are liquid or sticky they will adhere to any surface with which they come into contact. A dry particle, however, may rebound when it hits a surface and remain airborne until it reaches the next stage. Therefore, interstage losses would be less for a particle that bounces easily than for a sticky particle.

A further drawback of the Andersen cascade impactor is that it does not provide data in real-time. The procedure is slow and unsuitable for routine quality control purposes as the collection surfaces must be removed from the device and subjected to chemical or gravimetric analysis. The effect of humidity is also of significant importance when using the cascade impactor for testing dry powder inhalers. At relative humidities greater than 75%, adhesion of particles to the plates is good but below this limit, loss of particles is more evident (Hickey, 1992). Particles increase in weight with rising humidity and thus the cut-off diameter will become higher, yet the number of particles per stage remains the same at values of humidity that are less than 70%.

6.5.5 Other *in vitro* devices

Other *in vitro* models have also been employed for the characterisation of medicinal aerosols. These models are designed to mimic either inhalation profiles generated by patients or the geometry the respiratory tract. One of the earliest models, the 'Kirk lung', resembled the human trachea and bronchi in shape and size (Kirk, 1972). The two bronchi joined and proceeded to an adaptor holding a membrane filter capable of retaining particles with a diameter of 1 μm or greater. It was designed to operate at a flow rate of $16 \, l \, min^{-1}$. The 'Kirk lung' was later connected to a cascade impinger and operated at a range of flow rates up to $80 \, l \, min^{-1}$ (Davies *et al.*, 1976). Using this model, it was observed that the amount of powder deposited in the mouthpiece and the throat regions decreased with increasing flow rate.

Another model was developed to accommodate the operating requirements of an Andersen cascade impactor (e.g. flow rate of $28.3 \, l \, min^{-1}$) and at the same time, provide a flow rate representative of the asthmatic patient ($56.6 \, l \, min^{-1}$ at the mouthpiece) (Martin *et al.*, 1988). This model took into account the dimensions of the upper airways and control of temperature and relative humidity which was unique compared to other models.

The 'Electronic Lung™', an inhalation simulator was designed to characterise aerosol particle size over a wide range of pressure drops across the device rather than over a range of flow rates (Brindley *et al.*, 1993). In this model, a computer controlled piston draws air through the inhalation device according to a user-defined profile of pressure drop versus time. The aerosol generated passes through a large holding chamber and into a cascade impactor at a constant flow rate of $28 \, l \, min^{-1}$.

The twin impinger (TI), multistage stage liquid impinger (MSLI) and Andersen Cascade Impactor (ACI) are all recommended to be used for testing aerodynamic particle size distribution of aerosols by the British Pharmacopoeia and European Pharmacopoeia. However, they are required to be operated at different flow rates. Since drug delivery from dry powder inhaler is a function of inspiratory flow rate, these testing devices will give different aerodynamic particle size distributions for the same drug formulation. This is particularly problematic when pharmaceutical equivalence is being assessed. Thus, considerable harmonization of the testing methodologies is expected to be imminent.

REFERENCES

BOEHEM, G., KRUPP, H., LANGE, H. and SANDSTEDE, G. (1962) *Trans. Inst. Chem. Engrs. (London)* **40**, 18.

BOOTH, S.W. and NEWTON, J.M. (1987) Experimental investigation of adhesion between powders and surfaces. *J. Pharm. Pharmacol.* **39**, 679–684.

BRINDLEY, A., MARRIOTT, R.J., SUMBY, B.S. and SMITH, I.J. (1993) The Electronic Lung — A novel tool for the *in-vitro* characterization of inhalation devices. *J. Pharm. Pharmacol.* **45** (Suppl.) p. 1135.

BRITISH PHARMACOPOEIA COMMISSION (1993) Pressurised inhalations: deposition of the emitted dose. *British Pharmacopoeia*, Vol. II. London, HMSO, A194–A196.

CHUNG, F.H. (1981) Imaging and analysis of airborne particulates. In: CHEREMISSINOFF, P.N. (ED.), *Air/Particulate Instrumentation and Analysis*. Ann Arbor, Ann Arbor Science Publishers, 89–117.

CORN, M. (1961) Adhesion of solid particles to solid surfaces II. *J. Air Poll. Control Assoc.* **11**, 566–575, 584.

CORN, M. (1966) Adhesion of particles. In: DAVIES, C.N. (ed.), *Aerosol Science*. London, Academic Press, 359–392.

DAVIES, P.J., HANLON, G.W. and MOLYNEUX, A.J. (1976) An investigation into the deposition of inhalation aerosol particles as a function of air flow rate in a modified 'Kirk Lung'. *J. Pharm. Pharmacol.* **28**, 908–911.

ENDO, Y., HASEBE, S. and KOUSAKA, Y. (1997) Dispersion of aggregates of fine powder by acceleration in an air stream and its application to the evaluation of adhesion between particles. *Powder Technol.* **91**, 25–30.

FUKUOKA, E., KIMURA, S., YAMAZAKI, M. and TANAKA, T. (1983) Cohesion of particulate solids. VI. Improvement of apparatus and application to measurement of cohesiveness at various levels of humidity. *Chem. Pharm. Bull.* **31**, 221–229.

GOTOH, K., KIDA, M. and MASUDA, H. (1994) Effect of particle diameter on removal of surface particles using high speed air jet. *Kagaku Kogaku Ronbun.* **20**, 693–700.

HALLWORTH, G.W. (1977) An improved design of powder inhaler. *Brit. J. Clin. Pharmacol.* **4**, 689–690.

HALLWORTH, G.W. and WESTMORELAND, D.G. (1987) The twin impinger: A simple device for assessing the delivery of drugs from metered dose pressurized aerosol inhalers. *J. Pharm. Pharmacol.* **39**, 966–972.

HALLWORTH, G.W., CLOUGH, D., NEWNHAM, T. and ANDREWS, U.G. (1978) A simple impinger device for rapid quality control of the particle size of inhalation aerosols delivered by pressurised aerosols and powder inhalers. *J. Pharm. Pharmacol.* **30**, 39.

HICKEY, A.J. (1992) Methods of aerosol particle size characterisation. In: *Pharmaceutical Inhalation Aerosol Technology*. New York, Marcel Dekker, 219–253.

HINDS, W.C. (1982) *Aerosol Technology: Properties, Behavior, and Measurement of Airborne Particles*. New York, John Wiley & Sons.

HOWE, P.G., BENTON, D.P. and PUDDINGTON, I.E. (1955) Interaction forces between particles in suspensions of glass spheres in organic liquid media. *Can. J. Chem.* **33**, 1375–1379.

CHAPTER 6

HUGOSSON, S., LINDBERG, J., LOOF, T. and OLSSON, B. (1993) Proposals for standardized testing of powder preparations for inhalation. *Pharm. Forum* **19**, 5458–5466.

JAEGFELDT, H., ANDERSSON, J.A.R., TROFAST, E. and WETTERLIN, K.I.L. (1987) Particle size distribution from different modifications of Turbohaler®. In: NEWMAN, S.P., MOREN, F. and CROMPTON, G.K. (eds), *A New Concept in Inhalation Therapy: Proceedings of an International Workshop on a New Inhaler*, 21–22 May, London, UK, Medicom, 90–99.

JASHNANI, R.N., BYRON, P.R. and DALBY, R.N. (1995) Testing of dry powder aerosol formulations in different environmental conditions. *Int. J. Pharm.* **113**, 123–130.

JENIKE, A.W. (1964) Storage and flow of solids. *Bulletin of the University of Utah*, Utah Engineering Experimental Station, 53.

KIRK, W.F. (1972) *In-vitro* method of comparing clouds produced from inhalation aerosols for efficiency in penetration of airways. *J. Pharm. Sci.* **62**, 262–264.

KOUSAKA, Y. (1991) In: IINOYA, K., GOTOH, K. and HIGASHITANI, K. (eds), *Powder Technology Handbook*. New York, Marcel Dekker, 417.

KRUPP, H. (1967) Particle adhesion theory and experiment. In: OVERBEEK, J.T.G., PRINS, W. and ZETTLEMOYER, A.C. (eds), *Advances in Colloid and Interface Science*. Amsterdam, Elsevier, 111–240.

LAM, K.K. and NEWTON, J.M. (1991) Investigation of applied compression on the adhesion of powders to a substrate surface. *Powder Technol.* **65**, 167–175.

LEE, M.G., IRELAND, D.S., WEIR, P.J. and DWYER, P.J. (1993) Comparison of salbutamol inhalers available in the United Kingdom. *Int. J. Pharm. Practice.* **2**, 172–175.

LIPPMANN, M. (1989) Sampling aerosols by filtration. In: HERING S.V. (ed.), *Air Sampling Instruments*, American Conference of Governmental Industrial Hygienists, 305–336,

MARPLE, V.A. and WILLEKE, K. (1979) Inertial impactors. In: LUNDGREN, D.A. (ed.), *Aerosol Measurement*. Gainesville, University Presses of Florida, 90–107.

MARTIN, G.P., BELL, A.E. and MARRIOTT, C. (1988) An *in-vitro* method of assessing particle deposition from metered pressurised aerosols and dry powder inhalers. *Int. J. Pharm.* **44**, 57–63.

MARTONEN, T.B. and KATZ, I. (1993) Deposition patterns of poly disperse aerosols within human lungs. *J. Aerosol Med.* **6**, 251–274.

MARTONEN, T.B, KATZ, I., FULTS, K. and HICKEY, A.J. (1992) Use of analytically defined estimates of aerosol respirable fraction to predict lung deposition patterns. *Pharmaceut. Res.* **9**, 1634–1639.

MEAKIN B.J. and STROUD, N. (1983) An evaluation of some metered dose aerosols using a twin stage impinger sampling device. *J. Pharm. Pharmacol.* **35** (Suppl.) p. 7.

MILLER, N.C., MARPLE, V.A., SCHULTZ, R.K. and POON, W.S. (1992) Assessment of the twin impinger for size measurement of metered-dose inhaler sprays. *Pharmaceut. Res.* **9**, 1123–1127.

MITCHELL, J.P., COSTA, P.A. and WATERS, S. (1988) An assessment of an Andersen Mark II cascade impactor. *J. Aerosol Sci.* **19**, 213–221.

PADFIELD, J.M., WINTERBORN, J.K., POVER, G.M. and TATTERSFIELD, A. (1983) Correlation between inertial impaction performance and clinical performance of a bronchodilator aerosol. *J. Pharm. Pharmacol.* **35** (Suppl.) p. 10.

PODCZECK, F., NEWTON, J.M. and JAMES, M.B. (1994) Assessment of adhesion and autoadhesion forces between particles and surfaces. 1. The investigation of autoadhesion phenomena of salmeterol xinafoate and lactose monohydrate particles using compacted powder surfaces. *J. Adhes. Sci. Technol.* **8**, 1459–1472.

PODCZECK, F., NEWTON, J.M. and JAMES, M.B. (1995) Assessment of adhesion and autoadhesion forces between particles and surfaces. Part II. The investigation of adhesion phenomena of salmeterol xinafoate and lactose monohydrate particles in particle-on-particle and particle-on-surface contact. *J. Adhes. Sci. Technol.* **9**, 475–486.

TILLERY, M.I. (1979) Aerosol centrifuges. In: LUNDGREN, D.A. *et al.* (eds), *Aerosol Measurement.* Gainesville, University Presses of Florida, 3–23.

VIDGREN, P., SILVASTI, M., VIDGREN, M., PARONEN, P., TUKIAINEN, H. and LEHTI, H. (1991) *In vitro* inhalation behaviour and therapeutic response of salbutamol particles administered from two metered dose aerosols. *Pharmazie* **46**, 41–43.

ZIMON, A.D. (1982) Adhesion, molecular interaction and surface roughness. In: *Adhesion of Dust and Powder*, 2nd edn. London, Consultant Bureau, 46–47.

CHAPTER 6

Physical Characterisation of Pharmaceutical Powders

Contents

7.1 INTRODUCTION

As has been emphasised in the preceding chapters, the performance of a pharmaceutical powder aerosol is a function of many of its physical properties such as the particle morphology (size, shape and surface texture), crystallography, water content and surface energy, to name but a few. It is well known that some subtle change in physical properties may result in a dramatic change in powder performance. Thus, it is critical to have a full understanding of these physical properties during preformulation studies in order to obtain not only the effectiveness, but also the consistency and uniformity of the delivery system. The need for physical characterisation becomes even more crucial when an excipient, such as lactose, is used in the formulation. The addition of the excipient will undoubtedly pose extra complications in comparison to the use of bulk drug powders alone. To avoid any possible problems during drug formulation development, the physical characterisation of bulk drugs, excipients, and their blends should become normal practice in the preformulation stage. Failure to study some of the subtle physical properties of the material may be one of major causes of the lack of understanding of particle formation processes, prediction of powder performance, batch to batch variation, particle interactions and the overall performance of powders (York, 1994).

A systematic approach to the physical characterisation of pharmaceutical solids has been outlined (Brittain et al., 1991), where physical properties are classified as being associated with the molecular level, particulate level or bulk level. Molecular properties of a powder are determined by the chemical structure of the component particles. The measurement of molecular properties of a small ensemble of individual molecules may theoretically represent the molecular properties of the bulk powder that is composed of only one specific material, or one that is uniformly mixed if two or more components are present. The properties which are important at the particulate level are defined as those characteristics that can be determined by the analysis of a small number of particles. Some typical particulate properties are particle size, particle shape and surface texture, etc. Bulk properties refer to those characteristics of a powder that require a large ensemble of particles for the measurement to be made, typical of which is powder flowability. This chapter will adopt this classification of the various properties of powdered materials when discussing some of the most widely employed techniques in the characterisation of pharmaceutical powders. Extensive coverage of this field is beyond the scope of this book so special emphasis will be placed upon the characterisation of the physical properties of powders employed in the formulation of dry powder aerosols. The physical characterisation of other pharmaceutical powders has been reviewed elsewhere (e.g. Brittain, 1995).

7.2   MOLECULAR PROPERTIES

Dry powder aerosols are composed of either micronised drug particles alone or drug particles adhered to coarser carrier particles as discussed in chapter 3. Lactose, the most commonly employed carrier, is known to have different crystal forms, namely, α-monohydrate, α-anhydrous and β-lactose and it also exists in the amorphous form. Under normal conditions, lactose α-monohydrate is the most stable and common crystal form used by the pharmaceutical industry although others such as β-lactose and α-lactose anhydrous are also commercially available. Different crystal forms of a material can display radically different solubilities and this in turn affects the dissolution and bioavailability characteristics of the compound. In addition, the chemical stability of one crystal form may also be different from another. The presence of a trace amount of other crystal forms is sometimes sufficient to cause a significant change in the physical properties. For example, a small amount of α-anhydrous lactose, which is highly hygroscopic, increases the moisture uptake of the bulk powder of lactose α-monohydrate. Similarly, the conversion of even a very small amount of β-lactose to α-lactose, which is thermodynamically more stable, may result in a profound increase in interparticulate forces due to solid bridge formation. Further, many drugs, especially steroids, are polymorphic so that after mechanical micronisation or solvent crystallisation, drug particles may contain some metastable or unstable crystal forms. Together with any amorphous drug particles, these unstable particles will transform to more stable crystal forms during storage which may result in a marked change in physical properties. Therefore, a full characterisation of the polymorphic properties of a material is critical in the formulation development process and it is very important to understand fully the crystallographic properties of both drug and excipients as well as their mixtures.

Further, different conditions of crystallisation may result in the production of a crystal with altered amorphous content. Such amorphous regions if they are on the surface of crystalline particles may lead to a significant change in the physico-chemical properties of the bulk powder. Therefore, apart from ensuring that the desired crystal form is present, the degree of crystallinity should also be controlled. Both crystal form and the degree of crystallinity of pharmaceutical solids can be determined using X-ray powder diffraction, thermal analysis, vibrational spectroscopy and NMR spectrometry

7.2.1   X-ray powder diffractometry

When crystalline solids are exposed to incident X-ray beams, the radiation is scattered in all directions. In some directions, the scattered beams are completely in phase and reinforce one another to form diffracted beams, a phenomenon that can be described by

CHAPTER 7

Bragg's law. If a perfectly parallel and monochromatic X-ray beam, of wavelength λ, is incident on a crystalline sample at an angle θ, then diffraction will occur if:

$$n\lambda = 2d \sin \theta \tag{7.1}$$

where d is the distance between the planes in the crystal and n is the order of reflection.

X-ray powder diffractometry (XRPD) is a well-established method used to analyse the crystallography of particles. The X-ray powder pattern of every crystalline form of a compound is unique, making this technique particularly suited to the identification of different polymorphic forms of a compound. It can also be used to identify solvated and non-solvated forms of a compound, if the crystal lattices of the two are different. Variable temperature XRPD is another important technique where the sample is placed on a stage which can be heated to the desired temperature. The method is extremely useful in the study of thermally-induced phenomena and can be considered to be complementary to thermal methods of analysis. Detailed theories of XRPD are beyond the scope of this text but these have received excellent coverage elsewhere (e.g. Nuffield, 1966).

The X-ray diffraction patterns of both lactose and salbutamol sulphate are shown in Figure 7.1 (Brittain *et al.*, 1991; Ticehursh *et al.*, 1994; Chawla *et al.*, 1994). It can be seen that the X-ray diffraction patterns of hydrous and Fast-Flo™ lactose are very similar, while anhydrous lactose exhibits a different pattern. These observations suggest that hydrous and Fast-Flo™ lactose have similar crystal forms whilst anhydrous lactose has a different internal crystal lattice. According to its X-ray diffraction patterns, micronised salbutamol sulphate is highly crystalline whereas when spray-dried, the drug powder clearly contains a significant amount of amorphous material since the X-ray diffraction pattern lacks distinctive peaks.

Although XRPD has been widely used to characterise crystal forms, the use of XRPD to examine pharmaceutical powders has many associated problems (Suryabarayanan, 1995). First, the unit cells of many organic compounds are large, when compared with inorganic compounds, leading to low accuracy of d-spacing data. Second, the unit cells of organic compounds are often of low symmetry, resulting in complex powder patterns. Third, the X-rays can penetrate deep into the specimen of many organic materials, leading to large beam transparency errors and finally, organic compounds are especially prone to preferred orientation. Therefore, XRPD is often used, in combination with other methods, such as thermal analysis, to characterise such powders fully.

7.2.2 Thermal analysis

Thermal analysis is a technique by which powder properties are determined as a function of an externally applied temperature (Wendlandt, 1986). It is particularly useful to establish compound purity, polymorphism, solvation, degradation, and excipient compatibility (Giron, 1986). It can detect endothermic processes including melting, boiling, sublimation, vaporisation, desolvation and solid–solid phase transition, chemical degradation as well as exothermic processes such as crystallisation and oxidation. This method is particularly useful in determining drug–excipient interactions.

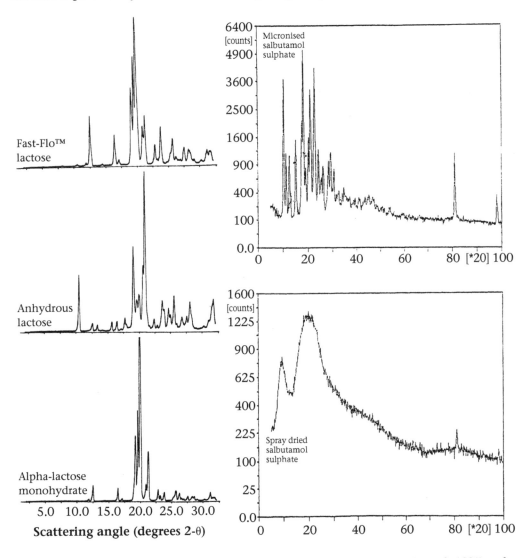

Figure 7.1: X-ray diffraction patterns of different crystal forms of lactose (Brittain *et al.*, 1991) and salbutamol sulphate (Chawla *et al.*, 1994).

CHAPTER 7

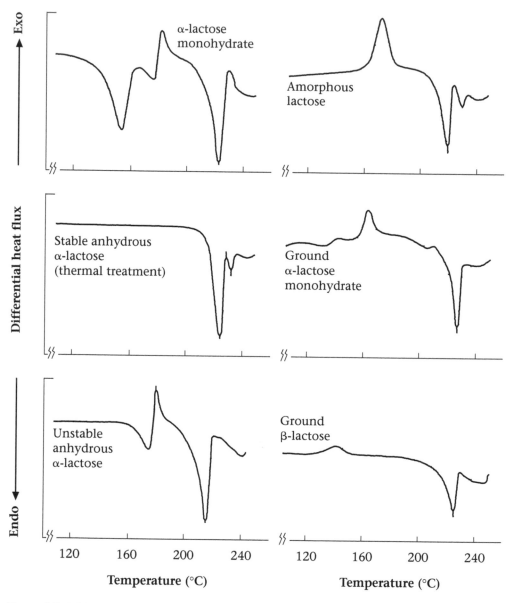

Figure 7.2: DSC thermograms of lactose (Lerk *et al.*, 1994a, b).

Thermal analysis encompasses the techniques of differential thermal analysis (DTA), differential scanning calorimetry (DSC) and thermogravimetry (TG). DTA monitors the difference in temperature between a sample and a reference as a function of external temperature whilst DSC measures the heat flow required to maintain the sample and reference at the same temperature. TG is a measure of the thermally-induced weight loss of a material as a function of the applied temperature. All three methods are now routinely used to characterise pharmaceutical solids and have been reviewed in depth elsewhere (McCauley and Brittain, 1995).

The DSC thermograms of some lactose and salbutamol sulphate particles are shown in Figures 7.2 and 7.3 respectively.

Thus, α-monohydrate lactose has an endothermic peak at about 140°C, corresponding to the loss of water of crystallisation (Figure 7.2). All three forms of α-lactose exhibit a melting endotherm commencing at about 210°C. The melting endotherm of unstable α-lactose was preceded by endo- and exothermic peaks at 170° and 180°C, respectively. The first endotherm at 170°C was thought to be due to the melting of the unstable anhydrous α-lactose, whereas the second endotherm at 210°C was attributed to the melting point of a crystalline β/α-lactose complex which crystallised at about 180°C. The amorphous lactose, prepared by spray-drying showed an exothermic transition at about 160°C and an endotherm at about 210°C. The DSC thermograms obtained for intensively ground α-monohydrate lactose and ground β-lactose were similar to that for amorphous lactose. These observations, taken together with the X-ray diffraction patterns, indicate that during extensive milling, α-monohydrate lactose loses its water of crystallisation and both α-lactose and β-lactose become more or less amorphous. After thermal treatment, amorphous lactose recrystallises. These observations are potentially of significance in the formulation of dry powders for inhalation not only because milled lactose is commonly used as a carrier but also since the change in the crystal form can occur during the micronisation process.

According to their DSC thermograms, both micronised and unmicronised salbutamol sulphate melt with degradation near 200°C (Figure 7.3). However, whilst the micronised sample showed a very small exothermic peak at 85°C, the unmicronised sample did not exhibit any significant thermal events prior to melting. By increasing the scale of the thermogram in the region of around 85°C, a small endotherm was noted at 64°C (Ward and Schultz, 1995). This endothermic peak was thought to correspond to the glass transition temperature of the amorphous material and it was found to disappear in a subsequent scan of the same sample. The exothermic peak at 85°C was consistent with a crystallisation event associated with the conversion of amorphous material to a crystalline state. This postulation was supported by the fact that following storage at 40°C/75% RH, the endotherm disappeared, possibly due to a prior amorphous to crystalline conversion in the humid environment.

7.2.3 Vibrational spectroscopy

Infrared spectroscopy (IR) is a fundamental technique that can be used in the identification of functional chemical groups of a molecule. Vibrational modes of a molecule are used to deduce structural information. These same vibrations can be used to study

CHAPTER 7

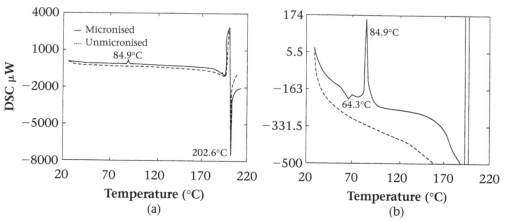

Figure 7.3: DSC thermograms of micronised and unmicronised salbutamol sulphate. Thermogram (b) was obtained with the instrument set at a higher sensitivity (Ward and Schultz, 1995).

solids, such as powders, since different solids having different internal three-dimensional structures display different vibrational modes in the IR spectra. However, the acquisition of high-quality infrared spectra on solid materials was made possible only with the advent of Fourier transform (FT) instruments. Fourier transform infrared spectroscopy (FTIR) has advantages over conventional IR in that it makes use of all the frequencies emitted by the source simultaneously and, therefore, provides an immediate increase in the signal-to-noise ratio (Markovich and Pidgeon, 1991).

Raman spectroscopy, another vibrational spectroscopic technique, is also valuable in the analysis of pharmaceutical solids. This technique involves irradiating the sample with a monochromatic laser beam and the inelastic scattering of the source energy is used to obtain a vibration spectra of the sample (Colthup *et al.*, 1975). IR and Raman spectroscopy are complementary to each other and when used together can provide more insight into the details of the structural composition of the compound, since the vibrational mode of some chemical groups may be infrared-active while others may be Raman-active. For example, asymmetric modes and vibrations due to polar groups (e.g. $C=O$, NH) are likely to be infrared-active, whereas symmetric modes and homopolar bonds (e.g. $C=C$) tend to be Raman-active. FTIR and Raman spectroscopy have been used in the identification of polymorphism, drug–excipient interactions, mixing and crystallinity (Bugay and Williams, 1995).

The near-IR spectra obtained for hydrous, anhydrous and Fast-Flo™ lactose are shown in Figure 7.4.

It can be seen that both hydrous and Fast-Flo™ lactose exhibit a large absorption band due to water at $5168 \, cm^{-1}$, whereas the corresponding level of absorption of anhydrous lactose is small. Since the water absorption band is often very evident in a near-IR spectrum, this region is considered to be useful in the determination of the water content of a solid.

7.2.4   Nuclear magnetic resonance spectroscopy

Nuclear magnetic resonance (NMR) spectroscopy has been used for the structure eluci-
dation of compounds but has required that the drug be in the solution-phase. With
recent advances in NMR hardware and software, the acquisition of high-resolution,

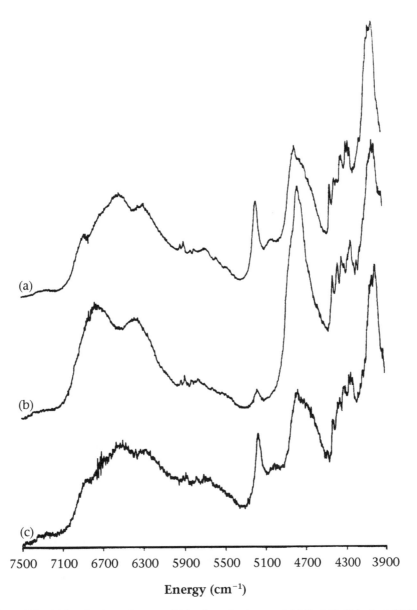

CHAPTER 7

Figure 7.4: FTIR of (a) anhydrous lactose, (b) hydrous lactose and (c) Fast-Flo™ lactose (Brittain *et
al.*, 1991).

multi-nuclear NMR spectra of compounds in the solid state has become possible and it is now being used in the characterisation of pharmaceutical solids (Bugay, 1993).

Conventional utilisation of solution-phase NMR data acquisition techniques on solid samples yield broad, featureless spectra, mainly due to dipolar interactions and chemical shift anisotropy, which can be averaged out to zero in the solution phase where molecular motion takes place randomly. This problem can be solved by spinning the sample at an angle of 54°44' (the so-called 'magic angle') with respect to the direction of the applied magnetic field. Even though the magic angle spinning (MAS) technique is able to obtain high-resolution ^{13}C NMR of materials in the solid state, the data acquisition time is lengthy due to the low sensitivity of the nuclei and the relatively long relaxation times. This can be further circumvented by the use of cross polarisation (CP), which transfers the spin polarisation from the high-abundance, high frequency nucleus (^{1}H) to the rarer low-frequency nucleus (^{13}C), allowing the build up of ^{13}C magnetisation and thus shortening the periods between pulses (Brittain et al., 1991).

The use of NMR in the characterisation of pharmaceutical solids, for example in the detection of polymorphism, is easily explained. If a compound can exist in different crystal forms, which are conformationally different, a carbon nucleus in one form may be situated in a slightly different molecular geometry compared with the same carbon nucleus in another form. The difference in the local environment will lead to a different chemical shift interaction for each carbon atom and ultimately a different isotropic chemical shift for the same carbon atom in different polymorphic forms. Thus, the solid state NMR spectra can reflect the conformational differences between different polymorphic forms and this technique is thus specifically useful in the characterisation of polymorphic forms. For example, similar to the IR spectra (Figure 7.4), the solid-state NMR spectra of Fast-Flo™ and hydrous lactose are essentially identical (Brittain, 1991) whereas anhydrous lactose shows a different spectrum (Figure 7.5), indicating that the Fast-Flo™ and hydrous lactose have similar (if not identical) crystal structures, but different to that of the anhydrous lactose.

There are a number of advantages in the use of ^{13}C CP/MAS NMR to characterise polymorphism (Bugay, 1995). In contrast to IR spectroscopy and X-ray diffraction, the intensity of the measured signal by solid state NMR is not affected by the particle size of the solids. Furthermore, signal intensity is directly proportional to the number of nuclei producing it and thus, by correct assignment of the NMR spectrum, the origin of the polymorphism can be inferred from the differences in the resonance position for identical nuclei in each polymorphic form. Finally, solid state NMR can be performed at any stage of powder processing such as blending, lyophilisation or tabletting. Therefore, it can provide a powerful tool for investigating the change in crystal form of a compound under different processing conditions.

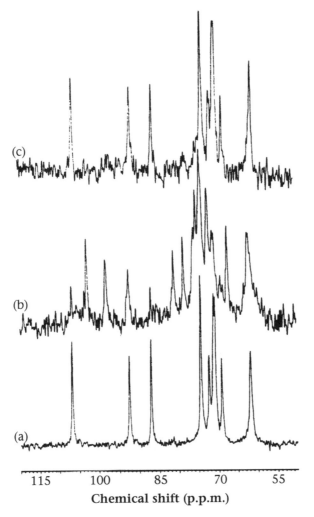

Figure 7.5: NMR spectra of (a) hydrous lactose, (b) anhydrous lactose and (c) Fast-flow™ lactose (Brittain *et al.*, 1991).

7.2.5   Miscellaneous techniques

UV/VIS diffuse reflectance spectroscopy is used to study the interactions of various compounds in a formulation, particularly those interactions that are characterised by a change in colour such as a change in intensity or type, which may result in coloration or discoloration. This method may be very useful for detecting any degradation of a solid material, which is associated with a change in colour. For example, the stability of ascorbic acid formulations has been successfully studied using diffuse reflectance spectroscopy, since degradation is accompanied by a colour change which is directly related to the potency of the active component (Carstensen *et al.*, 1964; Wortz, 1970).

7.3   PARTICULATE PROPERTIES

The particulate properties of powders are arbitrarily defined as those material character-istics which theoretically can be determined by the analysis of a small number of par-ticles. Any change in particulate properties will not affect the molecular properties of the compound but may significantly influence the bulk properties of the powders com-posed of these particles. Thus, the properties extant at the particulate level may include particle size, size distribution, shape and surface textures.

7.3.1   Particle size and size distribution

Particle size is the predominant factor governing the physical properties of a powder. The determination and control of particle size is often an essential requirement in the formulation of pharmaceuticals. This is particularly true in the formulation of dry powders for inhalation since the particle size of both carrier and drug particles deter-mines powder performance at every stage affecting the flowability, fluidisation, separa-tion and deposition of powders. The importance of particle size in the development of solid pharmaceutical dosage forms has long been appreciated and techniques for its measurement are well established. However, the particle size obtained is dependent upon the principle of measurement. A comprehensive coverage of the measurement and definition of particle size is outside the scope of this text, but these have been reviewed elsewhere (e.g. Edmundson, 1967; Allen, 1990).

The individual particles of a powder are rarely uniform in size and, therefore, it is often useful to define an average particle size and to use the associated parameters to compare samples. In describing the average diameter of a sample, three parameters can be used, namely, the mean, the median and the mode. The mean is the sum of the total number of particles divided by the number of particles. Since this is sensitive to extreme values, caution has to be employed in the use of the mean to describe the average parti-cle size. The median is the mid-value, 50% of the particles being distributed above the value and 50% below it, while the mode represents the size occurring most frequently. The median is influenced less by the size of extreme particles and is most commonly used to describe the average size of a powder.

Depending upon the mathematical equation used to assess particle diameter, the average particle size for a powder may be different. Table 7.1 lists some of the most com-monly used definitions of the mean diameter.

A number of mathematical functions have been proposed to represent the particle size distribution. The Gaussian, or normal distribution is given by:

$$y = \frac{1}{\sigma_n \sqrt{2\pi}} \exp\left[-\frac{(d - d_m)^2}{2\sigma_n^2}\right]$$ (7.2)

where y is the probability density (frequency), d is the diameter of a given particle, d_m is the arithmetic mean of the diameter of all the particles in the sample, and σ_n, the standard deviation, which is in turn defined by:

$$\sigma_n = \sqrt{\frac{\sum[n_i(d_i - d_m)]^2}{N}}$$ (7.3)

where N is the total number of particles in the sample and n_i the particles that have diameter d_i.

However, the Gaussian distribution is not common for powders produced by milling, grinding or mechanical micronisation. The distribution curve obtained is skewed and the frequency curve is then more likely to conform to the 'log normal' distribution, where d and σ_n in equation (7.2) are replaced by the logarithm of these quantities.

The standard deviation, σ_n, can be used to describe the particle size distribution for a normal distribution. The probability that a particle falls within $\pm 1\ \sigma_n$ is 68% whilst there is a 95% chance that it will fall within $\pm 2\ \sigma_n$. In a log-normal distribution, the logarithm of the particle size equivalent to 50% of the distribution is defined as the geometric mean diameter, d_g, which is also equal to the median diameter. The standard deviation is equal to the ratio of the 84.13% undersize divided by the 50% size.

A variety of methods have been employed to determine particle size and these include optical microscopy, laser light scattering, electrical zone sensing, sieve analysis and time of flight techniques. Each method is most appropriately used according to the intended application and each method may yield a different particle size function. For example, particle size measured by microscopy may be processed according to a number of different definitions as listed in Table 7.1, whilst the sizes measured with a Coulter counter are given as volume–number mean diameter, d_{vn}.

Optical or electron microscopy have proved to be powerful methods of size measurement, combined with some form of image analysis (Kaye, 1981). In addition to particle size, such techniques can also provide details about shape, crystal habit, surface texture and other information. However, there are some limitations to the use of microscopy in particle size determination. First, microscopy is a subjective measure that is prone to observer bias and error. In order to minimise this error, a sufficient number of particles should be counted randomly and this can be very time-consuming and tedious although this drawback has been partly overcome by the use of automatic image analysis techniques. Second, particles are examined in their plane of maximum stability and the equivalent particle size is derived in terms of two dimensions. This will often result in an overestimation of the particle size, especially for highly anisometric particles since the particles preferentially rest on the most stable plane. Third, the resolution of

CHAPTER 7

TABLE 7.1

Different definitions of particle size

Type of mean	Size parameter	Frequency	Mathematical equation	Description
Arithmetic	Length	Number	$\Sigma nd/\Sigma n$	Length–number mean, d_{ln}
	Surface	Number	$(\Sigma nd^2/\Sigma n)^{1/2}$	Surface–number mean, d_{sn}
	Volume	Number	$(\Sigma nd^3/\Sigma n)^{1/3}$	Volume–number mean, d_{vn}
	Length	Length	$\Sigma nd^2/\Sigma nd$	Surface–length mean, d_{sl}
	Length	Surface	$\Sigma nd^3/\Sigma nd^2$	Volume–surface mean, d_{vs}
	Length	Weight	$\Sigma nd^4/\Sigma nd^3$	Weight–moment, d_{wm}
Harmonic	1/length	Number	$(\Sigma nd^{-1}/\Sigma n)^{-1}$	Harmonic–number mean, d_{hn}
	Log length	Number	antilog $(\Sigma n \log d/\Sigma n)$	Geometric–number mean, d_{gn}
Geometric	Log length	Length	antilog $(\Sigma nd \log d/\Sigma nd)$	Geometric–length mean, d_{gl}
	Log length	Surface	antilog $(\Sigma nd^2 \log d/\Sigma nd^2)$	Geometric–surface mean, d_{gs}
	Log length	Weight	antilog $(\Sigma nd^3 \log d/\Sigma nd^3)$	Geometric-weight mean, d_{gw}

optical microscopes limits their application to the measurement of particles 1 μm or larger.

Sieve analysis provides another simple and direct method for the determination of particle size. Particles are allowed to pass through a series of screens by means of mechanical shaking or air-jet dispersion and the amount of material retained on each screen is determined. The data are most commonly displayed as the percentage of material retained on each sieve, the cumulative percentage of sample retained on each sieve or the percentage of the sample passing through each sieve. Apart from particle size analysis, sieving may also be useful in other stages of solid dosage form development. For example, a sieve classification scheme was employed to determine the presence of agglomerates formed during the mixing of fine particles with coarser carrier particles (Malmqvist and Nystrom, 1984a). Increasing the mixing time was found to decrease the amount of large agglomerates and indeed as mixing time was lengthened further such agglomerates disappeared finally, indicating the formation of uniform ordered mixes. A similar method was used to assess the effect of scale-up on mixing, and it showed that the mixing time required to obtain a uniform mixture was reduced substantially with increasing batch size (Malmqvist and Nystrom, 1984b).

The particle size obtained by sieving represents the minimum square aperture through which the particle can pass. Particles will pass through the openings on the screen according to their cross-sectional diameter rather than length. Therefore, it is likely that the sieved particles have true diameters larger than that calculated from the openings of the sieve through which the particles pass. This is particularly true in the case of elongated particles. Under extensive sieving, the constant change in the orientation of the particles may favour the passage of the particles via their shortest axes. Consequently, the shaking and vibration conditions must be optimised so as to ensure adequate separation of particles and to minimise particle-fracturing. Due to significant cohesive forces of smaller particles, sieve analysis is only appropriate for particles in excess of 38 μm (Randall, 1995), although small-mesh sieves and electrodeposited screens are under development to measure particles below this size. In general, sieving is preferentially limited to particles >75 μm.

If a non-conducting particle is suspended in a conducting medium and allowed to pass alone between two electrodes, the resistance between the electrodes will be momentarily increased and the magnitude of the increase in the resistance is related to the diameter of the particle. This is the basis of electrical zone sensing methods, sometimes referred to as the Coulter principle, used to determine particle size. When a particle of diameter d is suspended in an electrolyte solution and the suspension flows through a small orifice or aperture of diameter D with an immersed electrode on either side, the change in resistance ΔR is related to particle diameter, d, by:

$$\Delta R = \frac{53d^3}{D^4} \tag{7.4}$$

The change in the resistance creates an electrical pulse and the magnitude of the pulse size is proportional to the particle volume.

Electrical zone sensing, in an instrument such as a Coulter counter, is a well-established method for particle sizing. Applications in pharmaceutical science include the measurement of crystal growth in suspensions (Carless and Foster, 1966), the change in particle size of poorly water-soluble compounds during dissolution (Nystrom et al., 1985) and the counting of particulates in parenteral formulations (Stokes et al., 1975).

Particles to be measured by electrical zone sensing analysis must be easily dispersed in the electrolyte and must not readily agglomerate. It is best for the particles to have a narrow size distribution since too large a particle may block the aperture, while very small particles may not be detected. Depending upon the aperture size, the technique can be used to analyse particles in the range of about 0.4 to 1200 μm. Particle shape can also affect the results when the particles deviate significantly from sphericity, such as if the particles are presented as flakes. Another drawback to the Coulter method is that calibration using monodispersed particles of known diameter is required to assign the particle sizes of unknown species. Finally, particles that are soluble in the electrolyte medium are not suitable for electrical zone sensing analysis, unless the particle size determination is for a purpose such as monitoring crystallisation from the electrolyte solution or measuring dissolution into the medium.

Light scattering is another important particle size measurement. The observation that the maxima and minima in a shadow behind an obstacle are caused by interference waves, led to the exploitation of the phenomenon of light scattering more than 100 years ago. Many theories have been generated to study the light-scattering properties of particles, amongst which is the classical Mie theory. According to this theory, the

CHAPTER 7

scattering of light by particles approaching the wavelength of incident radiation is a function of the angles of scattering, the wavelength of the incident light, differences in refractive indices, and the diameter of a sphere. However, the mathematics of the Mie theory are so complicated that only after the advent of the computer was it possible to employ the theory to determine particle size. When particles are four or five times greater than the wavelength of the incident radiation, the Mie theory can be reduced to the simpler Fraunhofer diffraction theory, also referred to as static light scattering or low-angle forward light scattering. This latter theory relates the intensity of light scattered by a particle to the particle size, whereas the magnitude of the diffraction pattern is inversely proportional to the particle size. A more detailed treatment of the theory underlying light scattering can be found in various references (e.g. Van De Hulst, 1957; Kerker, 1969) and therefore will not be discussed further here.

One of the most widely used instruments based upon Fraunhofer theory is the Malvern Mastersizer (Malvern Instruments, Malvern, UK). The Fraunhofer diffraction-based instruments are applicable to the measurement of particles within the range of 1.2 to 1800 μm, depending on the lens used, although the standard working range for most instruments is roughly 2 to 300 μm (Frock, 1987). For particles under 5 μm, which are closer to the wavelength of the light, other light scattering techniques based on the combination of Fraunhofer and the more complex Mie theories has to be used, for example photon correlation spectroscopy (PCS) and dynamic light scattering. PCS is applicable for particles ranging from 0.001 to 5 μm and has been used widely to characterise many pharmaceutical particles including aerosols (Clay et al., 1983).

Light scattering is an absolute method that does not require calibration and is rapid to perform. Data derived from such methods can also provide information on size polydispersity. However, the particle size measured by light scattering may not correspond completely to that measured by other methods (Orumwense, 1992). Since they tend to be biased in favour of larger particles and the results are also affected by particle shape (Randall, 1995).

7.3.2 Particle shape

The shape of a particle is defined as the recognised pattern of relationships among all of the points which constitute the external surface (Luerkens et al., 1987). It is widely accepted that particle shape plays an important role in determining the physical properties of particles and numerous attempts have been made to quantify this parameter. However, no robust experimental methods have been developed that can quantitatively measure the shape of particles in a bulk powder, due to its complexity. Further, particle shape, as defined, is composed of two contributors, the geometric shape, that is the main dimensions of a particle's external surfaces, as well as the surface texture, which refers to the roughness of the surfaces. Most of the currently employed methods of shape analysis fail to distinguish between these two different aspects. In order to reveal the true shape of particles, a combination of various measurement techniques needs to be employed. Such complicated characterisation tends to reduce their applicability in routine analysis where direct and simple experimental methods are always favourable.

The various methods reported in the literature by which particle shapes may be classified can be divided into four main categories: literal description, assessment of individual particle characteristics, analysis by mathematical techniques and measurement of bulk properties of a powder.

7.3.2.1   Literal description

Particle shape is often habitually described by subjective descriptors, such as spherical, triangular, cuboidal, flake-like, fibrous, dendritic, etc. Non-objective statements are also used to express the extent of the shape of a particle. For example, particles have been described as being 'very spherical', 'not so round', 'rather elongated', 'somewhat smooth', 'more or less flat', etc. However, such descriptors are often very ambiguous and more robust methods, such as those employing mathematical equations, have to be used in order to describe more precisely the particle morphology.

7.3.2.2   Assessment of individual particle characteristics

The assessment of individual particle morphology usually takes the form of a comparison with standard geometric shapes, such as a circle or sphere. In a pioneering study, Heywood (1947) defined thickness, breadth and length as the three basic dimensions of an irregular particle resting on a plane in the position of its greatest stability (Figure 7.6). Thickness (T) was defined as the distance between two planes parallel to the plane of greatest stability and tangential to the surface of the particle, breadth (B) as the minimum distance between two parallel planes which are perpendicular to the plane defining thickness and are tangential to the opposite surfaces of the particles and length (L) as the distance between two parallel planes which are perpendicular to both those planes defining thickness and breadth and which are also tangential to the opposite surfaces of the particle.

Table 7.2 lists some of the most commonly used shape factors used to evaluate the deviation of a particle projected area from that of a sphere.

The elongation or flatness ratios may be the simplest shape factors derived from microscopic measurements. However, since such factors are derived from particles viewed in two-dimensions, it is not an adequate descriptor of three-dimensional particle shape. The aspect ratio is a factor dependent upon the orientation of a particle. It follows that, a sphere will have an aspect ratio of 1 whatever the orientation and particles of other shapes will exhibit aspect ratios of either <1 or >1. Both low and high values of the aspect ratio indicate an irregularity of a particle shape. Circularity, also termed roundness, is a factor combining both geometric shape and surface smoothness.

CHAPTER 7

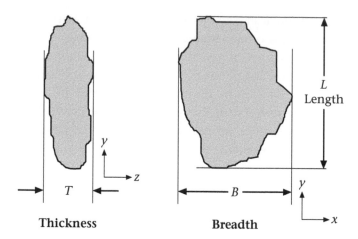

Figure 7.6: The measurement of particle thickness, breadth and length.

A sphere with surface asperities below the level of detection of the employed method of measurement will have a circularity value of 1. A sphere with rough surfaces and particles of any other shapes will possess circularity values >1. The higher the circularity value, the more irregular the shape of the particle. Sphericity, also termed rugosity (Carstensen, 1980), has been used to assess the surface smoothness of lactose particles (Kassem, 1990). However, sphericity is also a combination of particle geometric shape and surface smoothness. A sphere with negligible surface asperities will have a sphericity value of 1 but a sphere with detectable surface asperities and particles of any other shapes could also have sphericity values >1. Particles of the same geometric shape but with different surface smoothness will have different sphericity values and particles with the same surface smoothness but of different geometric shapes will also possess different sphericity values. Thus, sphericity can be used to compare particle surface smoothness only when the particles considered have the same geometric shape.

The shape coefficient is another common parameter used to determine particle shape. It is generated from the following equations (Singh, 1976):

$$A_\mathrm{p} = \alpha_\mathrm{s}\pi d^2 \tag{7.5}$$

$$V_\mathrm{p} = \alpha_\mathrm{v}nd^3 \tag{7.6}$$

where A_p and V_p are particle surface area and volume, respectively. d is the diameter of a sphere having the same projected area as that of the particle in its most stable orientation; α_s, the surface coefficient, and α_v, the volume coefficient, are both proportionality constants.

Thus, the surface-specific coefficient or shape coefficient K_a is calculated from equations (7.5) and (7.6):

$$K_a = \alpha_s/\alpha_v = A_p d/V_p \tag{7.7}$$

This factor is a measurement of three-dimensional particle shape and is also a combination of geometric shape and surface texture. At a constant volume, any means of increasing surface area including change in shape and surface smoothness will increase the value of the shape coefficient. If particles have smooth surfaces, the values of shape coefficient will increase from 6 for spherical particles to more than 100 for very thin flakes (Heywood, 1970). The higher the value of the shape coefficient, the more irregular the shape *or* the rougher the surfaces of the particle.

The shape factor, as defined below (Table 7.2), is a two-dimensional representation of particle shape. It is the shape parameter on which many image analysis techniques are based and once again, it is a combination of geometric shape and surface smoothness. The value of the shape factor for a perfectly smooth sphere is 1 and the value for any other shape will be less than 1, depending upon the geometric shape and surface rugosity of the particle. The lower the value of the shape factor, the more irregular the particle. However, values of the shape factor only vary between 0 and 1 and a large difference in particle morphology will be required to produce a marked change in its absolute value. This may be particularly the case when the particle morphology deviates distinguishably from a perfect sphere. Thus, the reciprocal of the value, i.e. $P_p^2/4\pi A_p$, may be a more appropriate factor to express particle shape since its values can vary from 1 to infinity.

Apart from the aforementioned shape factors, other factors have also been generated to measure particle shape under specific conditions. For example, the ratio of the

TABLE 7.2

Some most commonly used shape factors and their major properties

Shape factors	Definition	Value	Descriptions
Elongation ratio	L/B	$1 \to \infty$	Higher values indicate more elongated particles
Flatness ratio	B/T	$1 \to \infty$	Higher values indicate flatter particles
Aspect ratio	H_{max}/V_{max}	$0 \to \infty$	$1<$ or >1 for any shape except sphere
Circularity	P_p/P_c	$1 \to \infty$	>1 for any shape except circle or sphere
Sphericity	A_p/A_s	$1 \to \infty$	>1 for any shape except sphere
Shape coefficient	$A_p d/V_p$	$6 \to \infty$	Spheres have a value of 6
Shape factor	$4\pi A_p/P_p^2$	$0 \to 1$	Lower values indicate more irregular particle

P_p and P_c, perimeters of the projected image of the measured particle and a circle with the same area as the particle projected image, respectively; A_p and A_s, surface areas of the measured particles and a sphere having the same volume, respectively; V_p, particle volume; H_{max} and V_{max} are the maximum horizontal and vertical distances of the particle's projected image, respectively.

diameter of the internal contact sphere to that of the external contact sphere has been used to determine the shape of granules before compaction (Figure 7.7a) (Fukuoka and Kimura, 1992). This factor is able to distinguish between spherical and non-spherical particles but it has limited applications in the characterisation of particles that have small numbers of extremely deep gaps. According to its definition, it is a factor similar to the reciprocal of the elongation ratio. Yet another shape factor has been generated by dividing the actual projected area of a particle by the area of a circle having a diameter equivalent to the maximum projected length (Figure 7.7b). Such a factor has been employed to determine both geometric shape and surface smoothness of some powder systems (Otsuka *et al.*, 1988). In order to measure the sphericity of particles, Chapman *et al.* (1988) developed a method in terms of the angle necessary to tilt a plane such that the particles can be induced to roll. The method is based on the determination of the centre of gravity of the particle from a digitised image of the coordinates of its outline and computing the angle necessary to incline a plane such that the centre of gravity moves outside the boundary of the particle (Figure 7.7c). More spherical particles would have lower one-plane critical stability and thus require lower angles to roll. This method was said to be able to differentiate between small changes in roundness but it is a two-dimensional measurement and does not provide directly any other information on particle morphology. Further, a specialised computer system and program is needed and each particle has to be measured individually. A simplified shape factor, e_R, was devised to estimate the combination of surface asperities and deviation of shape from a circle (Podczeck and Newton, 1994) where e_R is defined by equation (7.8):

$$e_R = \frac{P_c}{P_p} - \sqrt{1 - \left(\frac{\text{breadth}}{\text{length}}\right)^2} \tag{7.8}$$

This equation takes into account both the deviation from circularity and the variation in surface roughness. It can give a value ranging from -1 to 1 and the smaller the value, the greater the shape irregularity.

None of the single shape factors mentioned above can distinguish between geometric shape and surface smoothness. Thus, a combination of these factors has to be employed

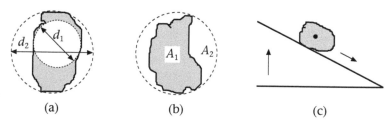

(a) (b) (c)

Figure 7.7: A schematic diagram showing how some shape factors are derived.

to differentiate between these two phenomena. Depending upon the purpose of measurement, different strategies may be employed. For example, the contribution of geometric shape to the overall sphericity value has to be eliminated when this factor is used to compare surface smoothness of different particles.

7.3.2.3 Analysis by mathematical techniques

Characterisation of particle shape can be carried out by using Fourier analysis (Luerkens et al., 1987). A two-dimensional closed curve is generated which approximates the perimeter of the projected image. Each point on the perimeter can be expressed by a set of (x, y) coordinates, a process known as digitising. These (x, y) pairs are then converted to polar coordinates, (R, θ) pairs, using the centre of gravity of the particle image as the origin at equally spaced sample points along the curve (Figure 7.8).

These (R, θ) pairs are then expanded into a Fourier series (Luerkens et al., 1982):

$$R(\theta) = a_0 + \sum_{n=1}^{N}(a_n \cos n\theta + b_n \sin n\theta) \tag{7.9}$$

where,

$$a_0 = \frac{1}{2\pi}\int_0^{2\pi} R(\theta)\, d\theta \tag{7.10}$$

$$a_n = \frac{1}{\pi}\int_0^{2\pi} R(\theta) \cos n\theta\, d\theta \tag{7.11}$$

$$b_n = \frac{1}{\pi}\int_0^{2\pi} R(\theta) \sin n\theta\, d\theta \tag{7.12}$$

and $n = 1, 2, 3, \ldots$

where a_0, a_n and b_n are the zero and nth order of the Fourier coefficients containing the size and shape information of the particle concerned. These can be employed to reconstruct the particle profile analytically using equation (7.9).

The Fourier coefficients are then transformed to a set of morphological descriptors such as size, shape, surface texture (Luerkens et al., 1987). These features measure well-defined properties of the particle profiles.

This technique has been shown to be able to differentiate between varying particle shapes (Yin et al., 1986). It is best suited to smooth, rounded particles but has limited application in characterising jagged or highly re-entrant surfaces (Clark, 1986). These problems can be solved by the use of a so-called fractal harmonics technique which is based upon fractal analysis. This technique has been used to quantify both the geometric shape and surface textures of particles. It is ideally suited to those with complex and convoluted surfaces (Young et al., 1990).

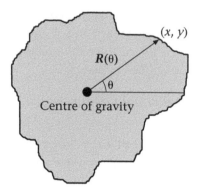

Figure 7.8: A schematic diagram showing the conversion of (x, y) coordinates to polar coordinates (R, θ).

7.3.2.4 Measurement of bulk properties

Some of the bulk properties of particles are affected by particle shape, (Beddow *et al.*, 1976) and many methods have been developed to determine the latter, based upon measurement of the bulk properties. Although the shape factors determined in this way are not representative of the shape profiles of individual particles, their values are physically relevant and can provide useful information despite often having a dependency on the technique used to obtain them. For example, the Heywood shape factor, which is the ratio of the function of particle surface area to the particle volume, can be determined practically by measurement of bulk properties such as powder surface area and density (Heywood, 1954). This shape factor has been widely used to derive an average particle shape for a bulk powder sample but it is not suitable for highly porous and re-entrant particles (e.g. Staniforth and Rees, 1981). However, the value of the Heywood shape factor is largely dependent upon the method employed to measure the surface area of the sample. For example, nitrogen adsorption measures every open-ended pore and irregularity in a powder particle and the surface area measured represents the total surface area. The corresponding shape factor will therefore have a disproportionately high value and thus represent an inflated value for shape since a portion of the measured surface area does not contribute to the external surface of the particle. Mercury intrusion porosimetry may be a better method to measure only the surface area that contributes to the morphology of particles, leaving the surface area contribution of the smaller pores unmeasured. The definition of 'small pores' will vary depending upon the process or use a powder is to be subjected to. For example, when undertaking a dissolution test, the smallest pores requiring consideration in the process will be those of the minimum size to allow penetration of water whereas when measuring the pores on a particle designed to act as a carrier in a dry powder inhaler, only those pores with a diameter greater than the smallest diameter of the drug particle needs to be taken into account. Thus, the value of the Heywood shape factor is a complex function of surface smoothness and geometric shape of the particle and for the same geometric shape, the rougher the surface, the higher the value of the factor. However, if the surface asperities are undetectable using a specific measurement, a sphere will have a Heywood shape factor of 6, whilst a smooth cube will have a value of 6.8 and more elongated and/or flaky particles will have even higher values.

The combination of the Heywood shape factor with the elongation ratio has been employed to produce another shape factor, α (Nikolakakis and Pilpel, 1985):

$$\alpha = \alpha_s/\alpha_v + N \qquad (7.13)$$

$$= S_w\rho d_H + N \qquad (7.14)$$

where α_s and α_v are the surface and volume coefficients, respectively; S_w ($m^2\,g^{-1}$) is the specific surface area, ρ ($g\,cm^{-3}$) is the particle density, d_H (μm) ($=(0.77 \times length \times breadth/\pi)^{1/2}$) is the mean Heywood equivalent diameter and N is the average elongation ratio.

This modified Heywood shape factor may be of more use in the determination of particle shape than the original form since it takes into account the geometric shape by including the elongation ratio. However, this factor is still unable to reveal the true morphology of particles adequately.

7.3.3 Surface smoothness

As discussed previously, surface smoothness plays an important role in determining interparticulate forces and the efficiency of removal of particles from either a stationary surface or the surfaces of airborne carrier particles in an air stream. The characterisation of the surface smoothness, particularly of the carrier particles, should form an important part of the formulation of dry powders for inhalation. Although the surface smoothness of the carrier particles with sizes of 63–90 μm may be best described qualitatively by optical or scanning electron microscopy, some mathematical calculations have to be employed in order to assess and compare surface smoothness quantitatively.

The surface roughness (rugosity) of lactose particles has been assessed by calculating the ratio of the surface area derived from air permeability and the calculated theoretical surface area, assuming all the particles were spherical (Kassem, 1990). The rugosity of commercial crystalline, re-crystallised and spray-dried lactose was reported to be 2.3, 1.2 and 2.6, respectively. However, such a rugosity factor, as defined, is the combination of both geometric shape and surface smoothness. If the surface asperities are negligible, the rugosity (sphericity) values should be 1 for a sphere, 1.24 for a cube and 3.76 for a regular tetrahedron. Thus, only when the contribution of geometric shape to the overall value can be ignored, should the factor be used to assess surface smoothness. However, due to the complexity of particle shape in a powder system, it is often impossible to generate single or combined mathematical equations that ignore the contribution of geometric shape to the overall rugosity value. Therefore, the characterisation of surface smoothness by means of such a rugosity factor is problematic.

A better method to determine the surface smoothness of particles may be the use of fractal geometry, which is related to the principle of self-similarity; i.e. the geometrical shape is kept identical independent of the scale, magnification, or power of resolution of measurement (Mandelbrot, 1982). A typical case of fractal geometry involves the measurement of the length of a wavy line. It is evident that the measured length of the line would increase with a decrease in the step length. Thus, the smaller the step length is, the longer is the line measured, since more directions of the line are taken into account. As shown in Figure 7.9, the measured perimeter of the particle is approximately 7.1 cm if a measuring step length of 1.02 cm is employed whereas measurement of the perimeter of the same particle is increased to about 9.7 cm if a step length of 0.36 cm is used.

Consequently, a direct relationship exists between the measured length of the line and the step length to measure it, depending upon the ruggedness or irregularity of the line. The more rugged or wavy the line, the greater dependence of the measured length on the step length. On the other hand, a straight line will have the same measured length, independent of the step length. Therefore, the ruggedness of a line can be calculated according to the relationship between the measured length of the line and the step length.

Similarly, the perimeter of the silhouette edge of a particle is dependent on the step length of measurement and the change in the measured perimeter with the step length can be used to characterise the surface roughness. This relationship is given by the following equations:

$$L_\delta = k\delta^{1-D} \qquad (7.15)$$

rearranging to give:

$$\log L_\delta = (1 - D)\log \delta + \log k \qquad (7.16)$$

where D is defined as the fractal dimension, δ denotes the step length and L_δ is the perimeter measured, k is a constant.

Thus, an ideal fractal structure should yield a linear plot at all resolutions, when $\log L_\delta$ is plotted against $\log \delta$, the slope of the line being $1 - D$. The fractal dimension, D, represents the degree of irregularity of the particle surface. The more irregular a substance is, the greater is the value of D (Farin and Avnir, 1987), so that $1 \leq D < 2$ for lines and $2 \leq D < 3$ for surfaces.

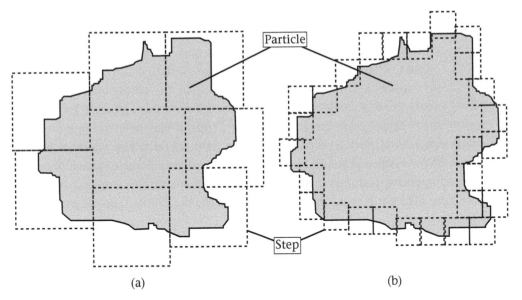

Figure 7.9: A diagram showing that decreasing step length increases the measured perimeter of a particle. (a) Step length 1.02 cm, seven steps; (b) step length 0.36 cm, 27 steps.

Fractal geometry has become a powerful tool for the characterisation of surface smoothness of pharmaceutical solids (e.g. Bergeron *et al.*, 1986; Farin and Avnir, 1992; Holgado *et al.*, 1995). For example, the fractal dimensions of some granular solids including lactose and sorbitol were shown to correlate well with the surface roughness of the particles when measured by scanning electron microscopy. Correlation was also found between the surface geometry and the physical behaviour of the granules such as flowability and compressibility (Thibert *et al.*, 1988).

The implementation of fractal analysis of boundary lines is expanding fast in many domains of the natural sciences, since the surfaces of many materials, either amorphous, crystalline, porous or non-porous are fractals and it has undoubtedly provided one of the most powerful tools for the physical characterisation of pharmaceutical particles. However, the silhouette edges of a particle are not the true image of the particle, and the use of the two-dimensional simplified profile still may not be able to reveal all the details of the three-dimensional morphology. However, a similar concept to fractal analysis may be applicable in the analysis of three-dimensional morphology with a view to obtaining a volume fractal. This may be particularly useful in the measurement of particle porosity since the void volume, as a function of power of resolution, should increase with increasing resolution of detection. If the relationship between the accessible void volume and the power of resolution of the pore size can be established, the total surface asperities of the particles can then be deduced. Since the penetration of mercury into surface openings is a function of mercury intrusion pressure, by carefully controlling the pressure, a predetermined accessible pore size can be obtained and the measuring resolution of this method can be changed from a lower to higher level. The total void volumes determined at different levels of resolution would be related to the overall surface asperities. This method has been used in the determination of tablet porosity (Leuenberger *et al.*, 1996) but its major limitation for this purpose is its inability to measure the occluded pores inside the particles. Thus, it may be even more useful in the characterisation of surface asperities of carrier particles for dry powders for inhalation, since only the open pores are significant in determining particle surface phenomena such as interparticulate forces and particle detachment. Future work is required in this particular field if it is to be able to predict the behaviour of dry powder inhaler formulations.

7.4 BULK PROPERTIES OF SOLIDS

7.4.1 Surface area

The surface area of a particulate solid can be defined in three different ways, namely, (a) as the visible or outer surface; (b) as the sum of the surface areas of all the particles which are regarded as non-porous units; (c) as the total area including fine structure and pores within the particles. In the pharmaceutical industry, surface area is becoming more and more important in the characterisation of materials during development, formulation and manufacturing. The surface area of a solid can provide vital information on the available void spaces on the surfaces of individual particles or a group of particles (Lowell and Shields, 1984). It is well known that many physico-chemical properties and even biological activities of some pharmaceutical solids may depend on the surface area of the solids. The measurement of the surface area of DPI formulations may also be important for the following reasons. First, the surface area may influence the interparticulate forces within the powder system. Second, it can be used to assess the particle morphology such as shape irregularity and surface asperities. Third, the pharmacokinetics of inhaled drug particles may be affected by their surface area and finally, some handling properties of materials, such as flowability, may also be related to surface area (Carstensen, 1993).

Many different methods are available for the determination of surface area of particulate systems with gas adsorption, mercury porosimetry and gas permeametry being the most widely used.

7.4.1.1 Gas adsorption

The adsorption of an inert gas such as nitrogen, helium, or krypton onto solid materials provides the most widely used method for surface area determination. The basic theory of this method is that the amount of gas adsorbed onto a particle is proportional to the surface area of that particle. The amount of the adsorbed gas can be calculated by measurement of its volume and pressure. In other words, the method involves the determination of the quantity of adsorbate (gas) required to cover all the available surface areas with a layer one molecule thick. The volume (V_m) of gas required to achieve monolayer adsorption at equilibrium under a specific temperature and pressure can be calculated according to the Langmuir equation:

$$\frac{P}{V} = \frac{1}{bV_m} + \frac{P}{V_m} \tag{7.17}$$

where V is the volume of the gas adsorbed at pressure P and b is a constant.

Thus, a plot of P/V against P will yield a straight line with slope $1/V_m$.

However, it is unlikely that the adsorbed gas will be limited to a monolayer and multimolecular layers are often formed on the surface caused by the condensation of gas molecules onto the adsorbed monolayer. This problem was solved by Brunauer, Emmett and Teller, who extended the Langmuir theory to account for multimolecular layer adsorption. The equation they generated is the classic BET equation (Brunauer *et al.*, 1938):

$$\frac{P}{V(P_0 - P)} = \frac{1}{V_m C} + \frac{(C - 1)P}{V_m C P_0} \tag{7.18}$$

where V is the volume of gas adsorbed at pressure P (the partial pressure of the adsorbate), V_m is the volume of gas adsorbed as a monolayer, P_0 is the saturation pressure of adsorbate at the experimental temperature and C is a constant exponentially relating the heats of adsorption and condensation of the adsorbate.

Therefore, using various concentrations of adsorbate, a graph of $P/V(P_0 - P)$ against P/P_0 yields a straight line. The value of V_m can be calculated from the reciprocal of the sum of the slope and the intercept. Then, the total surface area of the sample powder is calculated using:

$$S_t = \frac{V_m N_0 A_{cs}}{M_w} \tag{7.19}$$

where S_t is the total surface area, N_0 is Avogadro's number, A_{cs} is the cross-sectional area of the adsorbent molecule and M_w is the molecular weight of the adsorbent.

The specific surface area (S) of the solid can then be obtained from:

$$S = \frac{S_t}{m} \tag{7.20}$$

where m is the mass of powder measured.

The BET theory was based upon many assumptions such as all adsorption sites are energetically equivalent and the heat of adsorption for the second layer and above are

■ CHAPTER 7 ■

equal to the heat of liquefaction. Although such assumptions are not strictly true under normal conditions, this method has nevertheless proven to be an accurate representation of surface area for a wide range of materials examined (Newman, 1995). Nitrogen has been found to be an appropriate adsorbant gas for most materials which have surface areas of more than $1.0 \, \text{m}^2 \, \text{g}^{-1}$, while krypton should be used for those with smaller areas.

Nitrogen adsorption may also be used to determine the surface areas of both the carrier and drug particles of a DPI formulation. However, the surface area of carrier particles determined by nitrogen adsorption will consist of a portion of the surface that is not significant as far as physical interaction between the drug and carrier particles is concerned. It would be reasonable to assume that this part of the surface should be ignored in order to obtain a more accurate correlation between the surface area and particle behaviour.

7.4.1.2 Permeatric methods

Permeatry has also been widely used in the determination of surface area. If a fluid is allowed to flow through a powder bed under the influence of a pressure drop, the principle resistance to the fluid is created by the particle surfaces since the permeability of a powder bed is an inverse function of its surface area.

The streamline flow of a fluid (gas or liquid) through a bed of powder is related to the area of the fluid–solid interface. The surface area of the powder can be estimated using the Kozeny–Carman equation (Kaye, 1967) if an incompressible fluid flows through an homogeneous bed of smooth, non-porous particles:

$$S_v = \sqrt{\frac{\Delta P t A \epsilon^3}{kLV\eta(1-\epsilon)^2}}$$

(7.21)

where S_v is the volume-specific surface, ΔP is the pressure drop across the length, L, of the powder bed, A is the cross-sectional area of the bed normal to the flow direction, V is the volume of fluid flowing in time t, η is the viscosity of the fluid, ϵ is the powder bed porosity and k is a constant dependent upon particle shape and roughness, usually around 5 for spherical particles.

The above equation is however based upon some assumptions (Alderborn et al., 1985) and many difficulties arise with porous materials or materials with large surface areas (Singh, 1976). However, this approach has found considerable application in powder technology because of the simplicity of the apparatus and the rapidity of the measurement. It is generally more convenient to use a gas rather than a liquid and air permeability is normally adopted for routine measurements (e.g. Casal et al., 1989). The surface areas of a range of pharmaceutical solids measured by an air permeatry technique and microscopy were found to correlate well, although slightly higher surface area values were obtained using the former (Ericksson et al., 1990).

7.4.1.3 Mercury porosimetry

Mercury porosimetry has been one of the most widely used methods to characterise the interstitial voids of a solid (Hayes and Rossi-Doria, 1985). Mercury behaves as a non-wetting liquid for most substances and will not penetrate the solid unless pressure is applied. Mercury porosimetry is based on the relationship between the applied pressure and the minimum diameter of an open pore that can be filled with mercury. For a circular pore, the Washburn (1921) equation relates the applied pressure and the radius of the pores that can be intruded with a non-wetting liquid:

$$r = \frac{-2\gamma \cos \theta}{P} \tag{7.22}$$

where r is the pore radius, γ the surface tension of the liquid, θ the contact angle between the liquid and the sample and P the applied pressure.

Since mercury has a surface tension of 480 dynes cm^{-1} and a contact angle of 140° towards most substances, equation (7.22) can be then reduced to:

$$r = \frac{106.7}{P} \tag{7.23}$$

The inverse relationship between pore radius and pressure indicates that lower pressures are needed to measure large pores and vice versa. If mercury is forced into a powder bed, increasing applied pressure would increase the volume of mercury intruded into the powder. A plot of the volume of mercury intruded versus applied pressure can thus be obtained. On the other hand, if the external pressure is reduced mercury will extrude from the pores and this will cause a reduction in the intruded volume of mercury and an extrusion curve can then be generated. However, the intrusion and extrusion curves are unlikely to be the same and the resultant hysteresis loop is due to some mercury being permanently trapped in the pores (Lowell and Shield, 1984).

Many parameters can be derived from porosimetry data, such as the average pore radius (Orr, 1969/1970), surface area (Rootare and Prenzlow, 1967; Van Brakel et al., 1981), pore surface area (Lowell and Shields, 1984) and even particle size (Orr, 1969/1970). However, there are a number of limitations associated with the use of mercury porosimetry. First, the Washburn equation assumes that the pores are cylindrical in shape but this may not accurately represent the real situation. Second, deformation of the powder sample may occur at high applied pressures, especially with highly porous particles. Third, the contact angle and surface tension of mercury under the specific conditions should be determined in order to obtain the most accurate measurement. Finally, similar to gas adsorption and permeatry, mercury porosimetry can only measure the open pores of the samples. Some of these limitations may not pose major problems in the case of the characterisation of particles to be included in DPI formulations. Both drug and carrier particles should ideally be crystalline, and hence, the existence of closed pores is rare and deformation under normal conditions of

CHAPTER 7

measurement is unlikely to occur. Therefore, mercury porosimetry, combined with other methods of surface characterisation, may provide valuable information about the surface structures of both drug and carrier particles and this in turn is likely to provide a more detailed insight into particle properties relating to interparticulate forces, deaggregation, dispersion within the air stream, etc.

7.4.2 Flowability

The flowability of powders is such an important property that nearly all handling processes are affected to a greater or lesser extent. In the case of dry powder aerosol formulations, flowability of both drug and carrier particles plays an important role in mixing, capsule-filling and possibly aerosolisation. Flowability of powders is, for example, a function of particle size, size distribution, particle shape, interparticulate forces and surface textures (Neumann, 1967). Therefore, appropriate measurement of particle flowability forms an indispensable part of formulating dry powders for inhalation. Many methods have been documented to measure powder flowability, which include flow through an orifice, angle of repose, compressibility and shear cell methods (Amidon, 1995).

7.4.2.1 Flow through an orifice

The flow of non-cohesive powders under the influence of gravity has found many applications in industrial processes. Measurement of powder flow through an orifice is one of the most direct means of measuring powder rheology. Powders are allowed to flow through an orifice of a specific diameter for a predetermined period of time. The weight of powder passing through the orifice is recorded and the flow rate is expressed in terms of the quantity of powder flowing in a unit period of time. It is obvious that the higher the flow rate the more flowable the particles. The flow rate of powder through an orifice is a function of the diameter of the orifice, D, particle diameter, d, particle shape and density (Neumann, 1967). This simple and straightforward method has been widely employed in formulation development to assess powder flowability, although it is only applicable to free-flowing materials. It cannot be used for highly cohesive powders since the orifice may become occluded or the flow may follow a pulsating pattern.

The relationship between particle diameter and the flow rate through an orifice is shown in Figure 7.10. Fine particles may flow poorly due to cohesive forces whilst excessively large particles may also exhibit a poor flowability due to wall effects. The flow of a cohesive, monodisperse powder becomes poorer the smaller the particle diameter. Below a particular diameter, flow will not take place at all. The flow rate of particles of similar size, shape and density through the orifice of the same diameter can be employed to compare the cohesiveness of mixed powdered materials.

Highly cohesive powders may be mixed with a coarser powder so as to improve flow

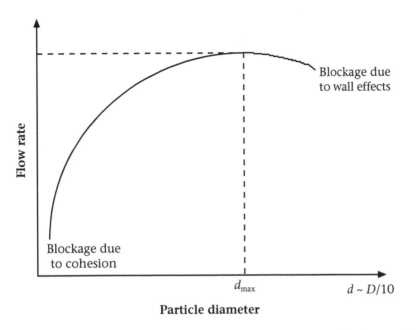

Figure 7.10: Flow rate through an orifice (diameter D) as a function of particle diameter (d).

through an orifice. The amount of the coarser powder required for the fine powder to exhibit a particular flow rate can be used to estimate the cohesiveness of the powder sample. For example, the higher the amount of coarse particles needed to achieve a certain flow rate, the more sticky or cohesive the fine particles.

7.4.2.2   Angle of repose

The angle of repose is the maximum angle possible between the surface of a pile of powder and the horizontal plane. It is by far the most commonly used method for the characterisation of powder flowability. The angle of repose of a pile of powder is not strongly dependent on average particle size, but more on those properties of the surface that determine powder flow properties (Neumann, 1967). The tangent of the angle of repose is equal to the coefficient of friction, μ, between the particles, which is largely dependent upon the surface roughness of the particles. However, although the angle of repose is primarily a function of surface roughness (Fonner et al., 1966) particle shape may also affect the value of this angle (Ridgway and Rupp, 1969). Its exact value also depends on the method of measurement (Neumann, 1967). Some experimental variations such as powder-bed consolidation, particle segregation, and aeration in forming the powder cone may reduce its reproducibility and reliability in representing flowability of particles. Poor correlation between the angle of repose and other methods such as flow through an orifice has been observed (Danish and Parrot, 1968). Another drawback is the low sensitivity of the method. Measured values of angle of repose only range from about 27° to 45° for powders. The theoretical minimum for uniform spheres, calculated from their geometry, is about 20°. Similar angles of repose may be observed for powders that have distinguishably different flow rates through an orifice (Amidon, 1995). Therefore, the angle of repose should be used in combination with other methods in order to provide a reliable characterisation of powder flow.

Associated with the angle of repose is 'the angle of spatula', which is defined as the angle formed when material is raised on a flat surface out of a bulk pile. This angle has also been used to estimate powder flowability (Carr, 1965). The test is carried out by forming a heap of the powder, placing a flat spatula into the bottom of the mass and then lifting it straight up and out of the material. The angle of the new surface on the spatula to the horizontal is measured immediately and again after gentle tapping of the spatula. The average of the two measurements is taken as the value of the 'angle of spatula'. It is generally larger than the angle of repose since the latter is formed after the powder heap has settled into a more stable arrangement.

7.4.2.3 Compressibility

Compressibility of a powder is usually expressed by the so-called compressibility index, C_i,

$$C_i = \frac{\text{Initial volume} - \text{Final volume}}{\text{Initial volume}} \times 100 \tag{7.24}$$

where the initial volume is the volume of the powder bed before tapping whilst the final volume is the volume of powder bed after tapping for a certain number of times.

Particle size, shape, surface texture, moisture content and cohesiveness of materials all affect the compressibility of a powder bed (Carr, 1965). In recent years, the compressibility index has become a simple, fast and popular method of powder flow measurement. A plot of C_i against the number of tappings will reveal valuable information about the flowability of a powder. The C_i will increase with increasing number of tappings until a plateau is reached. Higher values of C_i indicate poorer flowability of a powder which will need more tappings to reach equilibrium.

7.4.2.4 Shear properties of powders

The cohesion and flow properties of packed powders may be defined more precisely by their shear strength or resistance to shearing under different normal loads, and this has been described in chapter 6 of this book.

7.4.3 Density

Density is defined as the ratio of the mass of an object to its volume. A powder may have different densities, depending upon the definition of the volume of the powder.

The bulk volume, V_b, is the volume composed of the total space occupied by all the component particles, pore volume within the particles and the void spaces between the particles (Figure 7.11a). When the powder sample is tapped, the interparticle space decreases but the tapped volume, V_{tap}, is still composed of some interparticle void volume apart from the total space occupied by all the component particles and the pore volume within the particles (Figure 7.11b). In the case of particle volume, V_p, the interparticle void space is not considered and hence V_p only consists of the pore volume within the particles and the volume of the component particles (Figure 7.11c). The true volume, V_{true}, is the measurement of the real volume the component particles occupy within a powder (Figure 7.11d).

Therefore, these volumes follow a decreasing order of $V_b > V_{tap} > V_p > V_{true}$. Accordingly, the density of a powder bed may be expressed in terms of bulk density, ρ_b; tap density, ρ_{tap}; particle density, ρ_p and true density, ρ_{true} (Table 7.3), which increases in the order $\rho_b < \rho_{tap} < \rho_p < \rho_{true}$.

As can be seen from Table 7.3, bulk density is obtained by dividing the mass of a powder by the bulk volume of the powder bed (Mukaida, 1981). This is also referred to as apparent density (Mukaida, 1981). Tap density is the ratio of powder mass to the tap volume and may also be referred to as the drop density. Particle density or granular density (Mukaida, 1981) is calculated by dividing the mass of particles by the particle volume, which can be measured by the displacement of mercury (only pores greater than 10 μm being considered). For non-porous solids, particle density is similar to the true density of the powder, the latter parameter being obtained by dividing the powder mass by the true volume of the powder.

The most common means of measuring bulk density is to pour the powder into a tared graduated cylinder and the bulk volume and mass of the powder are then measured. In order to ensure the reproducibility of the measurement, a standard procedure has been introduced. This involves passing a sample of 50 g through a US Standard No. 20 sieve and then pouring the sample into a 100-ml graduated cylinder. The

■
CHAPTER 7
■

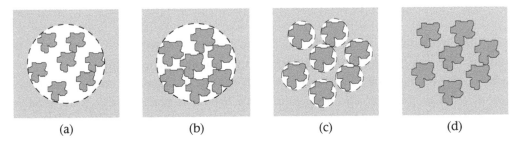

Figure 7.11: Different types of volumes of a powder (a) Bulk volume, V_b; (b) tap volume, V_{tap}; (c) particle volume, V_p; (d) true volume, V_{true}.

TABLE 7.3

Definitions and functions of different densities of a powder with mass, W

Densities	Definitions	Volume	Applications
Bulk density	W/V_b	Measured directly without excessive vibration	Powder packing, flow and compaction properties
Tap density	W/V_{tap}	Measured after tapping powder in a container	Powder packing, flow and compaction properties
Particle density	W/V_p	Measured by mercury displacement	Particle porosity
True density	W/V_{true}	Measured by gas pycnometry	Arrangement of atoms or molecules, crystalline nature

cylinder is then dropped from a height of 1 inch onto a hard surface three times at 2-s intervals. The volume of the powder is then read and used to calculate the bulk density.

Measurement of tap density also varies according to the protocol used and this is mainly due to the number of taps used for the measurement. The powder is placed in a cylinder, which is then tapped for a predetermined number of times before the volume is read. The powder may be tapped for 200, or 500, or even 1000 times (Doelling and Nash, 1992). It is obvious that different tap numbers will result in different tap volumes. Powders of different physico-chemical properties may require different optimum tap numbers, which must be established experimentally.

Particle volume can be derived from the porosity of the particles. The most common method of measuring porosity is by mercury intrusion, which has been dealt with earlier (section 7.4.1.3). The true volume of a powder can be measured by gas pycnometry, which is based on the ideal gas law, when a known quantity of gas at pressure, P_1, is allowed to flow from a predetermined volume, V_R, into a calibrated sample cell with a volume V_c containing the solid. The pressure within the reference and sample cells will quickly reach an equilibrium, P_2, and the sample volume can be calculated from the following equation:

$$V_{true} = V_c - V_R \left(\frac{P_1}{P_2} - 1 \right) \qquad (7.25)$$

From the true volume, the true density can be easily calculated if the powder mass is known. The gas employed to measure the true volume of a powder, must not be adsorbed by the material being measured. Both helium and nitrogen obey the ideal gas law at ambient pressures and temperatures but helium is preferred due to its smaller molecular size (Lowell and Shields, 1984).

Powder density has been found to correlate with a number of physical properties of the powder. For example, measurement of particle density may reflect the crystallinity of a powder since crystalline materials are usually more dense than amorphous materials

(Duncan-Hewitt and Grant, 1986). Bulk and tap densities are also representative of particle shape and more anisometric particles usually have smaller values since they tend to pack more loosely than more isometric particles (Neumann, 1967). Most importantly, powder density plays a critical role in determining powder flow and compaction properties. For example, powders of higher density usually have better flow properties (Harwood, 1971; Bos *et al.*, 1991). There are at least two major reasons for this. First, a higher density indicates that the particles are less porous and are therefore more likely to exhibit smaller interparticulate attractive and frictional forces when compared to more porous particles of the same material. Second, the gravitational forces due to particle mass, relative to interparticulate forces, may be more significant in the case of denser particles than lighter particles of a similar particle size. Therefore, high particle density favours free flow. The compressibility of a powder is also closely related to its density and the former can be estimated from the tap and bulk density measurement by equation (7.26) (Carr, 1965):

$$\text{compressibility} = \frac{\rho_{tap} - \rho_{bulk}}{\rho_{tap}} \times 100\% \qquad (7.26)$$

Therefore, a powder having a smaller bulk density has a higher compressibility value. An easily compressible material will be less flowable, and powders with compressibility values greater than 20% have been found to flow poorly (Carr, 1965). Such a phenomenon is obvious since compressibility of a powder is directly related to the deformability of its component particles. Higher deformability leads to stronger interparticulate forces and hence, poorer flowability of the powder.

Regulatory bodies require that appropriate acceptance criteria and tests should be instituted to control the properties of the drug and excipients which are essential to ensure their safety and efficacy. These include parameters derived from all three levels of particle characterisation such as specific identification, purity, impurity profiles, colour, appearance (visual and microscopic), moisture, residue on ignition, specific rotation, assay, residual solvent content, heavy metals, microbial limits, melting range, particle size distribution, surface area, crystalline form(s), particle shape, surface texture, etc. Other properties such as flowability and density should also be fully characterised during the development stage in order to produce a product that will perform satisfactorily.

The global initiative to eliminate CFC use and the need to develop efficient and user friendly inhaler devices has driven the development of CFC-free pMDIs, breath-actuated inhalers and dry powder inhalers. The pharmaceutical industries have made great efforts to refine the existing devices and to develop novel devices for the delivery of drugs for either localised or systemic effects. With the advancement in computing and information technology, more and more sophisticated inhaler devices will be made available to the patients. Further, more in-depth understandings are being gained in the

CHAPTER 7

fundamental properties of either liquid or solid aerosols, which is expected to lead to rapid development of novel techniques in both formulation design and particle engineering. Such a combined effort by device engineers and formulation scientists have continued into the 21st century and should enable the delivery of more products that will best suit the needs of the patients and extend the range of diseases that can be treated.

REFERENCES

ALDERBORN, G, DUBERG, M. and NYSTROM, C. (1985) Studies on direct compression of tablets. 10. Measurements of tablet surface–area by permeametry — reply. *Powder Technol.* **43**, 285–286.

ALLEN, T. (1990) *Particle Size Measurement,* 4th edn. New York, Chapman and Hall.

AMIDON, G.E. (1995) Physical and mechanical property characterization of powders. In: BRITTAIN, H.G. (ed.), *Physical Characterization of Pharmaceutical Solids.* New York, Marcel Dekker, 299–309.

BEDDOW, J.K., VETTER, A.F. and SISSONS, K. (1976) Powder metallurgy review (a) Part 1, Particle shape analysis. *Powder Metall. Int.* **8**, 69–70, 75–76.

BERGERON, M., LAURIN, P. and TAWASHI, R. (1986) Effects of particle morphology in selecting pharmaceutical excipients. *Drug Dev. Ind. Pharm.* **12**, 915–926.

BOS, C.E., VROMANS, H. and LERK, C.F. (1991) Lubricant sensitivity in relation to bulk density for granulations based on starch or cellulose. *Int. J. Pharm.* **67**, 39–49.

BRITTAIN, H.G. (1995) (ed.) *Physical Characterization of Pharmaceutical Solids.* New York, Marcel Dekker.

BRITTAIN, H.G., BOGDANOWICH, S.J., BUGAY, D.E., DEVINCENTIS, J., LEWEN G. and NEWMAN, A.W. (1991) Physical characterization of pharmaceutical solids. *Pharmaceut. Res.* **8**, 963–973.

BRUNAUER, S., EMMETT, P.H. and TELLER, E. (1938) Adsorption of gases in multimolecular layers. *J. Am. Chem. Soc.* **60**, 309–319.

BUGAY, D.E. (1993) Solid-state nuclear-magnetic-resonance spectroscopy — theory and pharmaceutical applications. *Pharmaceut. Res.* **10**, 317–327.

BUGAY, D.E. (1995) Magnetic resonance spectrometry. In: BRITTAIN, H.G. (ed.) *Physical Characterization of Pharmaceutical Solids.* New York, Marcel Dekker, 93–125.

BUGAY, D.E. and WILLIAMS, A.C. (1995) Vibrational spectroscopy. In: BRITTAIN, H.G. (ed.), *Physical Characterization of Pharmaceutical Solids.* New York, Marcel Dekker, 59–91.

CARLESS, J.E. and FOSTER, A.A. (1966) Accelerated crystal growth of sulphathiazole by temperature cycling. *J. Pharm. Pharmacol.* **18**, 697–708.

CARR, R.L. (1965) Evaluating flow properties of solids — classifying flow properties of solids. *Chem. Eng.* **72**, 163–168.

CARSTENSEN, J.T. (ed.) (1980) *Solid Pharmaceutics: Mechanical Properties and Rate Phenomena*. London, Academic Press, 34–36.

CARSTENSEN, J.T. (1993) *Pharmaceutical Principles of Solid Dosage Forms*. Lancaster, PA, Technomic Publishing Co.

CARSTENSEN, J.T., JOHNSON, J., VALENTINE, W. and VANCE, J. (1964) Extrapolation of appearance of tablets and powders from accelerated storage tests. *J. Pharm. Sci.* **53**, 1050–1054.

CASAL, J., ARNALDOS, J. and PELLICER, N. (1989) Measurement of the external surface area of powders by a permeametric method. *Powder Technol.* **58**, 93–97.

CHAPMAN, S.R., ROWE, R.C. and NEWTON, J.M. (1988) Characterization of the sphericity of particles by the one plane critical stability. *J. Pharm. Pharmacol.* **40**, 503–505.

CHAWLA, A., TAYLOR, K.M.G., NEWTON, J.M. and JOHNSON, M.C.R. (1994) Production of spray dried salbutamol sulphate for use in dry powder aerosol formulation. *Int. J. Pharm.* **108**, 233–240.

CLARK, N.N. (1986) Three techniques for implementing digital fractal analysis of particle-shape. *Powder Technol.* **46**, 45–52.

CLAY, M.M., PAVIA, D., NEWMAN, S.P. and CLARKE, S.W. (1983) Factors influencing the size distribution of aerosols from jet nebulizers. *Thorax* **38**, 755–759.

COLTHUP, N.B., DALY, L.H. and WIBERLEY, S.E. (1975) *Introduction to Infrared and Raman Spectroscopy*, 2nd edn. London, Academic Press.

DANISH, F.Q. and PARROT, E.L. (1968) *J. Pharm. Sci.* **49**, 92–96.

DOELLING, M.K. and NASH, R.A. (1992) The development of a microwave fluid-bed processor. 2. Drying performance and physical characteristics of typical pharmaceutical granulations. *Pharmaceut. Res.* **9**, 1493–1501.

DUNCAN-HEWITT, W.C. and GRANT, D.J. (1986) True density and thermal expansivity of pharmaceutical solids: comparison of methods and assessment of crystallinity. *Int. J. Pharm.* **28**, 75–84.

EDMUNDSON, I.C. (1967) Particle-size analysis. In: BEAN, H.S., BECKETT, A.H. and CARLESS, J.E. (eds), *Advances in Pharmaceutical Sciences*, Vol. 2. London, Academic Press, 95–149.

CHAPTER 7

ERICKSSON, M., NYSTROM, C. and ALDERBORN, G. (1990) Evaluation of a permeametry technique for surface area measurements of coarse particulate materials. *Int. J. Pharm.* **63**, 189–199.

FARIN, D. and AVNIR, D. (1987) Relative fractal surfaces. *J. Phys. Chem.* **91**, 5517–5521.

FARIN, D. and AVNIR, D. (1992) Use of fractal geometry to determine effects of surface morphology on drug dissolution. *J. Pharm. Sci.* **81**, 54–57.

FONNER, JR., D.E., BANKER, G.S. and SWARBRICK, J. (1966) Micromeritrics of granular pharmaceutical solids. II. Factors involved in sieving of pharmaceutical granules. *J. Pharm. Sci.* **55**, 576–580.

FROCK, H.N. (1987) In: PROVDER, T. (ed.), *Particle Size Distribution: Assessment and Characterization*. Washington, DC, American Chemical Society, 146–160.

FUKUOKA, E. and KIMURA, S. (1992) Cohesion of particulate solids. VII. influence of particle shape on compression by tapping. *Chem. Pham. Bull.* **40**(10), 2805–2809.

GIRON, D. (1986) Applications of thermal analysis in the pharmaceutical industry. *J. Pharmaceut. Biomed.* **4**, 755–770.

HARWOOD, C.F. (1971) Compaction effect on flow property indexes for powders. *J. Pharm. Sci.* **60**, 161–163.

HAYES, J.M. and ROSSI-DORIA, P. (1985) *Principles and Applications of Pore Structural Characterization*. Bristol, J.W. Arrowsmith.

HEYWOOD, H. (1947) The scope of particle size analysis and standardization. In: *Proceedings of the Symposium on Particle Size Analysis* Supplement to *Trans. Inst. Chem. Engr.* **25**, 14.

HEYWOOD, H. (1954) Particle shape coefficients. *J. Imp. Coll. Chem. Eng. Soc.* **8**, 25–33.

HEYWOOD, H. (1970) In: EVERETT, D.H. and OTTEWILL, R.H. (eds), *Proceedings of the International Symposium on Surface Area Determination*. London, Butterworths, 375.

HOLGADO, M.A., FERNANDEZ-HERVAS, M.J., FERNANDEZ-AREVALO, M. and RABASCO, M.A. (1995) Use of fractal dimensions in the study of excipients: application to the characterization of modified lactoses. *Int. J. Pharm.* **121**, 187–193.

KASSEM, N.M. (1990) Generation of deeply inspirable dry powders. PhD thesis, University of London.

KAYE, B.H. (1967) Permeability techniques for characterizing fine powders. *Powder Technol.* **1**, 11–22.

KAYE, B.H. (1981) *Direct Characterization of Fine Particles*. New York, John Wiley & Sons.

KERKER, M. (1969) *The Scattering of Light and other Electromagnetic Radiation.* New York, Academic Press.

LERK, C.F., ANDREAE, A.C. and DE BOER, A.H. (1994a) Alteration of lactose during differential scanning calorimetry. *J. Pharm. Sci.* **73**, 856–857.

LERK, C.F., ANDREAE, A.C. and DE BOER, A.H. (1994b) Transitions of lactose by mechanical and thermal treatment. *J. Pharm. Sci.* **73**, 857–858.

LEUENBERGER, H., LEU, R. and BONNY, J.-D. (1996) Application of percolation theory and fractal geometry to tablet compaction. In: ALDERBRON, G. and NYSTROM, C. (eds), *Pharmaceutical Powder Compaction Technology.* New York, Marcel Dekker, 137–141.

LOWELL, S.L. and SHIELDS, J.E. (1984) *Powder Surface Area and Porosity.* New York, Chapman and Hall.

LUERKENS, D. W., BEDDOW, J. K. and VETTER, A. F. (1982) Morphological Fourier descriptors. *Powder Technol.* **31**, 209–215.

LUERKENS, D.W., BEDDOW, J.K. and VETTER, A.F. (1987) Structure and morphology — the science of form applied to particle characterization. *Powder Technol.* **50**, 93–101.

MALMQVIST, K. and NYSTROM, C. (1984a) Studies on direct compression of tablets. 8. A sieve classification method for the determination of agglomerates and the distribution of fine particles in ordered mixing. *Acta Pharm. Suec.* **21**, 9–20.

MALMQVIST, K. and NYSTROM, C. (1984b) Studies on direct compression of tablets. 9. The effect of scaling-up on the preparation of ordered mixtures in double-cone mixers. *Acta Pharm. Suec.* **21**, 21–30.

MANDELBROT, B.B. (1982) *The Fractal Geometry of Nature.* San Francisco, Freeman.

MARKOVICH, R.J. and PIDGEON, C. (1991) Introduction to Fourier transform infrared spectroscopy and applications in the pharmaceutical sciences. *Pharmaceut. Res.* **8**, 663–675.

McCAULEY, J.M. and BRITTAIN, H.G. (1995) Thermal methods of analysis. In: BRITTAIN, H.G. (ed.), *Physical Characterization of Pharmaceutical Solids.* New York, Marcel Dekker, 224–251.

MUKAIDA, K.I. (1981) Density – Measurement of small porous particles by mercury porosimetry. *Powder Technol.* **29**, 99–107.

NEUMANN, B.S. (1967) The flow properties of powders. In: BEAN, H.S., BECKETT, A.H. and CARLESS, J.E. (eds). *Advances in Pharmaceutical Sciences*, vol. 2. London, Academic Press, 181–221.

CHAPTER 7

NEWMAN, A.W. (1995) In: BRITTAIN, H.G. (ed.), *Physical Characterization of Pharmaceutical Solids*. New York, Marcel Dekker, 260–264.

NIKOLAKAKIS, I. and PILPEL, N. (1985) Effects of particle-size and particle-shape on the tensile strengths of powders. *Powder Technol.* **45**, 79–82.

NUFFIELD, E.W. (1966) X-*Ray Diffraction Methods*. New York, John Wiley & Sons.

NYSTROM, C., MAZUR, J., BARNETT, M.I. and GLAZER, M. (1985) Physicochemical aspects of drug release. 1. Dissolution rate measurements of sparingly soluble compounds with the coulter-counter Model TAII. *J. Pharm. Pharmacol.* **37**, 217–221.

ORR, C. (1969/1970) Application of mercury penetration to materials analysis. *Powder Technol.* **3**, 117–123.

ORUMWENSE, O.A. (1992) The kinetics of fine grinding in an annular ball mill. *Powder Technol.* **73**, 101–108.

OTSUKA, A., IIDA, K., DANJO, K. and SUNADA, H. (1988) Measurement of the adhesive force between particles of powdered materials and a glass substrate by means of the impact separation method. III. Effect of particle shape and surface asperity. *Chem. Pharm. Bull.,* **36**, 741–749.

PODCZECK, F. and NEWTON, J.M. (1994) A shape factor to characterize the quality of spheroids. *J. Pharm. Pharmacol.* **46**, 82–85.

RANDALL, C.S. (1995) Particle size distribution. In: BRITTAIN, H.G. (ed.), *Physical Characterization of Pharmaceutical Solids*. New York, Marcel Dekker, 157–186.

RIDGWAY, K. and RUPP, R. (1969) The effect of particle shape on powder properties. *J. Pharm. Pharmacol.* Suppl. 21, Suppl. 30S.

ROOTARE, H.M. and PRENZLOW, C.F. (1967) Surface areas from mercury porosimeter measurements. *J. Phys. Chem.* **71**, 2733–2736.

SINGH, K.S.W. (1976) In: PARFITT, G.D. and SINGH, K.S.W. (eds), *Characterization of Powder Surface with Special Reference to Pigments and Fillers*. London, Academic Press, 4–9.

STANIFORTH, J.N. and REES, J.E. (1981) Shape classification of re-entrant particles II: Description of re-entrant and non-re-entrant particle shapes. *Powder Technol.* **28**, 9–16.

STOKES, T.F., SUMNER, E.D. and NEEDHAM, T.E. (1975) Particulate contamination and stability of three additives in 0.9% sodium chloride injection in plastic and glass large-volume container. *Am. J. Hosp. Pharm.* **32**, 821–826.

SURYABARAYANAN, R. (1995) X-ray powder diffractometry. In: BRITTAIN, G.H. (ed.), *Physical Characterization of Pharmaceutical Solids*. New York, Marcel Dekker, 187–221.

THIBERT, R., AKBARIEH, M. and TAWASHI, R. (1988) Application of fractal dimension to the study of the surface ruggedness of granular solids and excipients. *J. Pharm. Sci.* **77**, 724–726.

TICEHURST, M.D., ROWE, R.C. and YORK, P. (1994) Determination of the surface properties of two batches of salbutamol sulphate in inverse gas chromatography. *Int. J. Pharm.* **111**, 241–249.

VAN BRAKEL, J., MODRY, S. and SVATA, M. (1981) Mercury porosimetry — state of the art. *Powder Technol.* **29**, 1–12.

VAN DE HULST, H.C. (1957) *Light Scattering by Small Particles*. New York, John Wiley & Sons, 383–393.

WARD, G.H. and SCHULTZ, R.K. (1995) Process-induced crystallinity changes of albuterol sulphate and its effect on powder physical stability. *Pharmaceut. Res.* **12**, 773–779.

WASHBURN, E.W. (1921) A method of determining the distribution of pore sizes in a porous material. *Proceedings of the National Academy of Sciences USA* **7**, 115–116.

WENDLANDT, W.W. (1986) *Thermal Analysis*, 3rd edn. New York, John Wiley & Sons.

WORTZ, R.B. (1970) Color stability of ascorbic acid tablets measured by light reflectance. *J. Pharm. Sci.* **56**, 1169–1173.

YIN, M.-J., BEDDOW, J.K. and VETTER, A.F. (1986) Effects of particle shape on two-phase flow in pipes. *Powder Technol.* **46**, 53–60.

YORK, P. (1994) Powdered raw materials: characterising batch uniformity. In: BYRON, P.R., DALBY, R.N. and FARR, S.J. (eds), *Respiratory Drug Delivery IV, Program and Proceedings*. Virginia, USA. 83–91.

YOUNG, B.D., BRYSON, A.W. and VAN VLIET, B.M. (1990) An evaluation of the technique of polygonal harmonics for the characterisation of particle shape. *Powder Technol.* **63**, 157–168.

CHAPTER 7

Index